Field Propulsion System for Space Travel
Physics of Non-Conventional Propulsion Methods for Interstellar Travel

Edited and Authored by

Takaaki Musha

Technical Research & Development Institute, MOD

&

Yoshinari Minami

Advanced Science-Technology Research Organization

(Formerly NEC Space Development Division)

CONTENTS

FOREWORD

The book entitled Field Propulsion System for Space Travel provides a comprehensive and modern view of the principles of field propulsion, a rapidly growing subject of research. This excellent book certainly serves as a textbook for propulsion courses and as a reference for researchers.

Both Authors were among a group of specialists at the Advanced Space Propulsion Investigation Committee (ASPIC) that was organized under the auspices of the Japan Society for Aeronautical and Space Sciences in 1994. In March 1996, ASPIC issued a research report, published by the Japan Society for Aeronautical and Space Sciences.

The committee addressed all kinds of non-chemical space propulsion systems such as electric propulsion, laser propulsion, nuclear propulsion and solar sail in order to allow space exploration beyond Solar System. Further, as an expletive space propulsion system, other forms of propulsion were also investigated, that is, field propulsion system which utilizes a strain on space, zero-point energy in a vacuum, electro-gravitic effect, non-Newtonian gravitic effect predicted from the Einstein Theory of Gravity, and the terrestrial magnetism.

These new areas of investigation, and the results obtained so far, are included in this book, allowing a rapid and progressive study that will contribute to the reader view of its impact in our future. The general plan of the book made the several chapters as independent as possible, to enable the reader to choose his own order of study, and an endeavor has been made to join to every concept the name of its author.

I strongly recommend the book to teachers, researchers on advanced propulsion concepts, engineers, and members of the general public who want to think and be challenged by attempts to picture and suggest the future of propulsion. Certainly, some of the ideas described in this book will, ultimately, guide human beings to boldly explore what lies beyond our Solar System.

Prof. Mario J. Pinheiro
Department of Physics and Institute Plasmas and Nuclear Fusion
Instituto Superior Tecnico
Portugal

PREFACE

At the present stage of space propulsion technology, the only practical propulsion system is chemical propulsion system, which is based on the expulsion of a mass to induce a momentum thrust. Since the maximum speed is limited by the product of the gas effective exhaust velocity and the natural logarithm of mass ratio, its speed is too slow for the spaceship to achieve the interplanetary travel and interstellar travel. Thus, the breakthrough of propulsion method has been required until now. Instead of conventional chemical propulsion systems, field propulsion systems, which are based on the General Relativity Theory, the Quantum Field Theory and other exotic theories, have been proposed by many researchers to overcome the speed limit of the conventional space rocket. Field propulsion system is the concept of propulsion theory of spaceship not based on usual momentum thrust but based on momentum derived from an interaction of the spaceship with external fields (i.e. pressure thrust). Field propulsion system is propelled without mass expulsion. The propulsive force as a pressure thrust arises from the interaction of space-time around the spaceship and the spaceship itself; the spaceship is propelled against space-time structure.

The new propulsion physics for such means may or may not be discovered in the future. The space propulsion science behind such notions as warp drive and gravity-control is examined to show the connections between the known, the anomalies, and the visions; the physical range from means using the presently accepted physical paradigm to other more speculative means using "frontier physics". However, if new propulsion physics is discovered, a new class of technologies would result, revolutionizing spaceflight and enabling humanity to reach other habitable planets.

As is well known at present, it is impossible by utilizing existing technologies to reach the outer rim of the solar system. For the purpose to explore concepts that could someday enable interstellar travel, a radically advanced space propulsion concept as field propulsion that rely on physics outside the present paradigms must emerge in the near future. All such schemes are based on momentum derived from an interaction of the spaceship with external fields in order to satisfy conservation of momentum.

Throughout this book, we have discussed an overview of field propulsion system for the purpose of manned space flight such as interplanetary travel and interstellar travel, which cannot be achieved by the conventional chemical propulsion system.

In the future, mankind will reach for another solar system by utilizing field propulsion system which will be realized someday.

Although many kinds of field propulsion theories are introduced in this book, it should be appreciated, that the theories have not been yet to verify as correct; therefore they are only ideas. However, seeds in great variety are sowed for the future.

We wish this book will present good advice for the researchers and students to make further researches on the field propulsion system.

We sincerely wish to thank cooperators, who are good friends as well as colleagues, for all the hard work they put into this book.

Yoshinari Minami
Advanced Science-Technology Research Organization
Japan

Takaaki Musha
Technical Research & Development Institute, MOD
Japan

ACKNOWLEDGEMENT

The authors appreciate Paul Murad (ret. US Department of Defense), Salvatore Santoli (INT-International Nanobiological Testbed Ltd.), Nuno Santos (Oporto University), Gary V. Stephenson (Seculine Consulting) and Hiroyuki Mikami (ret. Ministry of Defense, Japan) for their help and suggestions, which greatly improved this book as well as removing some errors.

Field Propulsion System for Space Travel
Physics of Non-Conventional Propulsion Methods for Interstellar Travel

2

CHAPTER 1

Introduction: Travel to Space in 21st Century

Abstract: In this chapter, we describe the brief history of space travel by mankind, grand view of the Space Travel in 21st century regarding Solar system exploration and stellar system exploration, NASA research project, and also present the outlook of actualization technology for space travel.

INTRODUCTION

The dream of the space travel by rocket propulsion originated on the edge of 1903 has steadily expanded the range of the exploration to not only the moon but also the planets in the solar system by the predecessors' efforts. However, in the present day, only a chemical rocket propulsion system supports this technology for space exploration.

It all started with a landing to the moon by Apollo 11 of July 21, 1969 that manned space exploration by the human was initially achieved. Mankind finally engraved the first step of the brilliant history tat included manned space exploration to the moon which was the nearest to the Earth among 60 satellites in the solar system. The arrival time to the moon depends on the selected orbit; for example in the case of Apollo 12, it takes three days (for the outward trip 80 hours and 72 hours coming home). Although a small amount of navigation time (3 days) is not the problem, but in the case of travel more distantly to a remote planet from the Earth, it is an already dreamlike story because the speed is too low in current rocket technology.

The distance 384,400km (cislunar distance) is too short a distance if one sees it from a space scale. It can be said that traveling to the planet and the stars by the method of going to the moon is impossible due to the present rocket technical limitations. It is a fact that the human race has no possession of the space propulsion technology that can accelerate to high speed in a short time currently available. If by analogy we look at space as the ocean, the human race is obtaining only a technology advanced within the vicinity of the shore in the sands with the boat. If the speed of the spaceship increases marvelously, that is, the arrival time to a planet is several hours or about several weeks, it is considerably surmountable. As for the space development of the 21st century, if it is no epoch-making space transportation systems advancement, human race's area of involvement will be limited to the earth neighborhood indefinitely, and a new finding is not obtained. It is understood that it is important for manned space exploration to develop the necessary space transportation system that uses very high speeds.

FEASIBILITY OF SPACE TRAVEL

1) Solar System Exploration

When cruising to other planets, innumerable orbits can be considered that depend upon the propulsion system, but there is restriction at the launch time of a spaceship and it cannot leave when you would like. Although that depends on selected orbit, regarding Mars, the arrival time of a one way trip to Mars takes about six months to about 12 months, for example. Although the arrival time of this level does not become a problem so the time for a one-way trip on an unmanned spaceship, it will serve as a big barrier, if it is a human mission.

Although not impossible, there is danger for the crew in a spaceship and it is not an efficient means. There are many very difficult problems, such as influence of the space environment to include radiation on the human body during the space cruise, the water, food and oxygen supply for 2 years, and prolonged weightlessness (0G) effects, an urgent return at the time of the occurrence of an accident, and relief supplies from the Earth.

The origin of the problem is that manned Mars exploration takes long-term time where the cruise speed of a spaceship is too slow. The second astronomical velocity (11.2 km/s) that a rocket obtains for earth escape is slightly slow compared with the orbital speed (24 km/s) of Mars, and the orbital speed (30 km/s) of the Earth. This is due to the maximum speed of a rocket that is limited by the product of gas effective exhaust velocity and the natural logarithm of mass ratio (≈ 7). The speed beyond this cannot be theoretically larger than from the propulsion principle of a rocket based on momentum conservation law.

Concerning a chemical rocket which has multi-stage composition, about ten km/s of speed is a practical limit (See Fig. **1**). If the speed of a rocket is 1000 times faster compared with the speed of Mars or the Earth, then a straight line orbit can be attained. Whenever you like always, it can reach to the target planet in a shorter time without restriction of orbital calculation, a start time, and return time, so that it may operate similar to a car as it were.

Figure 1: Chemical rocket engine.

For this reason, research of an electric propulsion system (ion thruster, plasma thruster, arc jet thruster), a laser propulsion system, a nuclear propulsion system (heat nuclear fission propulsion, nuclear fusion propulsion, anti-matter propulsion), *etc.* different from a chemical rocket was briskly examined. However, since aforementioned propulsion principles may be considered as similar to a chemical rocket, the maximum speed is also restricted and sufficient thrust (acceleration) is not obtained.

In order to overthrow the wall of such space propulsion technology, research in field propulsion with a high speed and high acceleration is examined in the U.S (organized planning: NASA BPP), Europe (organized planning: ESA), and Japan (on an individual basis). The propulsion principle of field propulsion is based on pressure thrust without mass expulsion like momentum thrust. The pressure thrust in field propulsion refers to a reaction with space-time itself (*i.e.* vacuum). The propulsive force as a pressure thrust arises from the interaction of space-time around the spaceship and the spaceship itself, the latter being propelled against the space-time structure. The field propulsion principle consists in the exploitation of the action of the medium field induced by such interaction and is thus based on some concepts in modern physics found in General Relativity, Quantum Field Theory, Quantum Cosmology and Superstring Theory including D-brane to bring about the best propulsive performance. If it compares, it will be the way a person who rode on roller skates is pushed from the back by other persons, and progresses. Field propulsion is positioned as a propulsion system with which human beings can exploit the solar system and beyond.

2) Stellar System Exploration

As for the number of short-distance stellar systems, 63 fixed stellar systems exist in less than 18 light years from the solar system, and 814 fixed stellar systems exist in less than 50 light years from the solar system. As an example of a short-distance stellar system, Alpha Centauri is the nearest star from the solar system in 4.3 light years, and the star Sirius is near the seventh nearest star in 8.7 light years. On the other hand, as an example of a long distance stellar system, the Pleiades star cluster exists 410 light years distance and Cygnus exists 1800 light years distance.

In the near future, the next target of humanity is to complete exploration of the stellar system. Moreover, although the SETI project planned in each country until now is premised on the actual existence of extraterrestrial intelligent life. Contact with extraterrestrial live may be impossible due to the lack of a suitable navigation theory and its technology which can conquer a huge distance between the stars and Earth. That is, even if the speed of light is obtained, we have to spend very long hours underway which will be required for tens of several year to hundreds years.

Considerable years are required even if it achieves interstellar travel at the speed of light.

To conquer huge distances and time, it is said that superluminal velocity becomes indispensable. Then, although the superluminal velocity tends to be expected simplistically, it becomes an unreal expectation regrettably from the basic theory of physics and restrictions on the propulsion theory.

There exists the wall created by the velocity of light due to special relativity, any propulsion principle cannot exceed the velocity of light, and there is no energy source as the power source accelerated to the wall of the velocity of light.

Special relativity theory acts correctly in actual space and any propulsion theory cannot exceed the wall of the velocity of light.

No propulsion theory but a new navigation theory becomes indispensable for stellar system exploration that requires the cruising range of a light-year unit. Although navigation by special relativity is well known, it is unreal navigation, which does not become useful for the extreme time gap of global time and spaceship time as is well known as the Urashima effect (twin or time paradox). Even if we could reach a fixed star in several years, what 100 years and what 1000 years had passed when it returned to the Earth. It literally becomes a one way space trip.

Therefore, research of space warp which uses the wormhole by the general relativity and the hyper-space navigation theory with the character of imaginary time are required to remove this Urashima effect.

3) NASA' Research Project

In the latest scientific research paper, a new theory and new phenomenon is discovered and the possibility of a space transportation system of a certain kind which can convey human beings to other star systems is suggested. With the appearance of these newest science and perceptions of impossibility of transportation to other star systems by using existing propulsion system, NASA established a research project named "Breakthrough Propulsion Physics" in December, 1996.

The greatest concern for research of NASA is the theory and the basic experiment about combination of gravity and an electromagnetic field, fluctuation of the vacuum zero-point energy, a warp drive, a wormhole, the quantum gravity effect, *etc.* Since the distance to a stellar system is so large, even travel to the fixed star nearest to the Earth using the present propulsion technology will require tens of thousands of years. To overcome such a limit of space travel between fixed stars, research and development on new propulsion theory, propulsion engineering, and the navigation theory are furthered by this project.

Actualization Technology for Space Travel

A field propulsion system is based on the propulsion principle of a certain kind of pressure thrust as compared to the rocket system based on momentum thrust (reaction thrust) by gas injection. Although the solar sail and light sail are also based on a kind of a solar pressure that produces thrust with optical pressure, it differs from the pressure thrust as used in field propulsion.

The propulsion principle of a field propulsion system is the concept of a system of using an interaction with the structure between the space-time continuum, making full use of the newest modern physics (the theory of relativity, quantum theory, cosmology, *etc.*), and propelling a spaceship within the structure of space-time. Although various kinds of field propulsion systems are proposed, the following common performances are obtained: the maximum theoretical speed is quasi speed of light, high acceleration (several G to several tens of G) and no inertial force.

Modern physics accepts vacuum zero-point energy, and this zero-point energy has a theoretical relationship in inertia and gravity. And the space, which is in a vacuum is a place that has repeated creation and annihilation of a particle and an antiparticle continuously, and which has physical substance. It is necessary to search for a mechanism that generates the acceleration field for exciting the vacuum asymmetrically locally to the vacuum with such substance. Advance to human beings' true universe may be carried out by realization of a new propulsion theory and navigation theory based on the newest modern physics.

The present chemical rocket propulsion technology is effective in human beings' moon or Mars advance, and effective also in outer planet exploration by an unmanned deep space probe. However, when it comes to the planetary exploration by humanity in suitable time, some propulsion physics of new substitution is required. Search of the alternative propulsion system by a physical principle that is different to overcome such a limit is needed, and it becomes important to systematically investigate such possibilities.

If the space development in the 21st century will not have an epoch-making advance of a space transportation system, mankind's domain will be forever limited near the Earth, and new knowledge will not be acquired.

The goal of a mankind's space travel in the 21st century needs to attain to beyond even the stellar system not only a solar system. By developing physics further, it is required to make the new propulsion technology of get ahead of the limit of existing propulsion system. The domain where the present physics is not yet completed is left, and it is expected that practical development of an epoch-making new propulsion system and the navigation theory will be achieved with the faster development of physics.

Finally we write down the next words of Robert Goddard (the developer of the world's first liquid fuel rocket) who gives hope and courage to a solitary researcher.

"Yesterday's dream is today's hope, and it is the tomorrow's reality". (Robert H. Goddard).

CHAPTER 2

Conventional Propulsion Technologies

Abstract: Besides the practical chemical propulsion system, many propulsion systems and concepts, such as electric propulsion system, laser propulsion system, nuclear propulsion system, solar sail, nuclear electric propulsion, and solar electric propulsion are introduced. In this chapter, the outline propulsion theory are presented.

INTRODUCTION

Nowadays, aside from the practical chemical propulsion system, many propulsion systems and concepts are investigated. Those are as follows: Electric propulsion system (ION thruster, MPD thruster, ARC jet thruster, and Hall thruster), Laser propulsion system, Nuclear propulsion system, Solar sail, NEP (nuclear electric propulsion), SEP (solar electric propulsion) [6].

In this chapter, the outline and its propulsion theory of electric propulsion system, laser propulsion system and nuclear propulsion system are introduced in the main.

ELECTRIC PROPULSION SYSTEM

Outline of Electric Propulsion

Electric propulsion behaves as rocket propulsion but is not chemical propulsion. Electric propulsion heats and ionizes a propellant by electric energy, then accelerates the propellant by various methods to generate the reaction thrust. Since the specific impulse (I_{SP}) of electric propulsion is large compared with chemical propulsion, the electric propulsion can reduce the amount of consumption of propellant. Therefore it allows the increase in payload weight and space missions which can not be attained by chemical propulsion. Electric propulsion is classified into electrostatic acceleration type (ion thruster), electric heat acceleration type (arc jet thruster), and electromagnetism acceleration type (plasma thruster) according to the difference in the thrust generation mechanism. An ion thruster is one of the former typical concept and arc jet engine, MPD, and a Hall thruster are one of the concepts represented by the latter.

(a) Ion Thruster

An ion thruster (e.g.Fig. **1**) is an equipment which transforms a propellant into ionization plasma (*i.e.* ion), and accelerates these ions to high velocity by electrostatic power to produce thrust. The thrust of an ion thruster is several mN to several 100 mN, a specific impulse is from 2000 seconds to 7000 seconds and the propulsion efficiency has the performance in the order of 80%. In space, only the positive charge ions cannot be emitted continuously. Since only the part to which the engine itself emitted ion is charged in a minus electric charge, the emitted ion is pulled back by electrostatic force, and then the thrust will decline. Moreover, an injection ion beam spreads by electrostatic repulsive force between ions and thrust reduction occurs with the increase of radial speed.

For this reason, the ion beam injected from the accelerator grid needs to neutralize nearby and the neutralizer assembly which emits the electrons of the same number into an ion beam are generally needed.

Moreover, there is a peculiar problem that thrust density cannot be enormous due to space charge limited current for an ion thruster. In general, a maximum value exists in the amount of electric charges which occupy the inside of a fixed volume and this is called space charge restrictions. Since the ion thruster cannot send bigger ion current than space charge limited current, thrust is restricted. In order to enlarge thrust, the diameter of an ion thruster will be also enlarged, but this has technical constraint according to the several conditions.

(b) Microwave Discharge Ion Thruster

The ion thruster needs to ionize a propellant and to generate the plasma as an ion source firstly and uses high frequency electric discharge for this plasma production. A microwave and a magnetic field are used together in order to utilize an Electron Cyclotron Resonance (ECR) for electric discharge of plasma production.

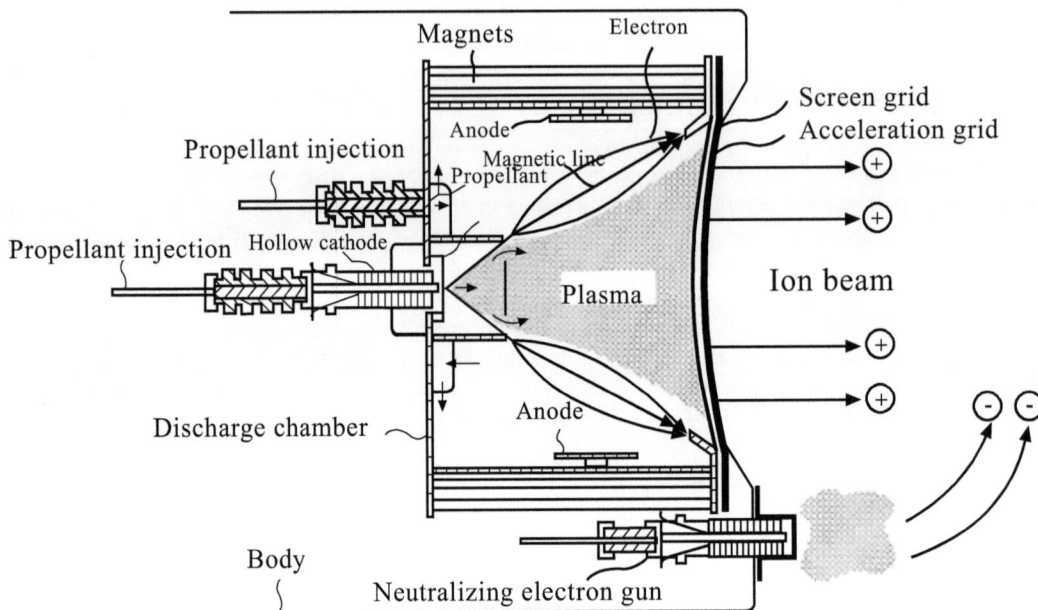

Figure 1: Kaufman type ion thruster

Generally, an electron is coiled and attached to a line of magnetic field, and performs spiral rotational movement. The rotation frequency (Rama frequency) in this rotational movement is called cyclotron frequency. By supplying from the outside the electromagnetic waves which are in agreement with this cyclotron frequency, an electron resonates with clockwise circular polarization electric field, absorbs energy and is accelerated. An electron is continuously accelerated in the ECR domain.

Fig. **2** shows a microwave electric discharge type ion thruster. It arranges so that a magnetic pole may become an electric discharge indoor part which turns around a ring-like permanent magnet and the cusp field is generated. It consists of an electric discharge room, a microwave generator for ion source generation of an electric discharge room, a neutralizer assembly, a source generation neutralization of electron dexterous microwave generator, a propellant tank and a grid of three sheets. The injection of microwave can perform plasma ignition in an instant.

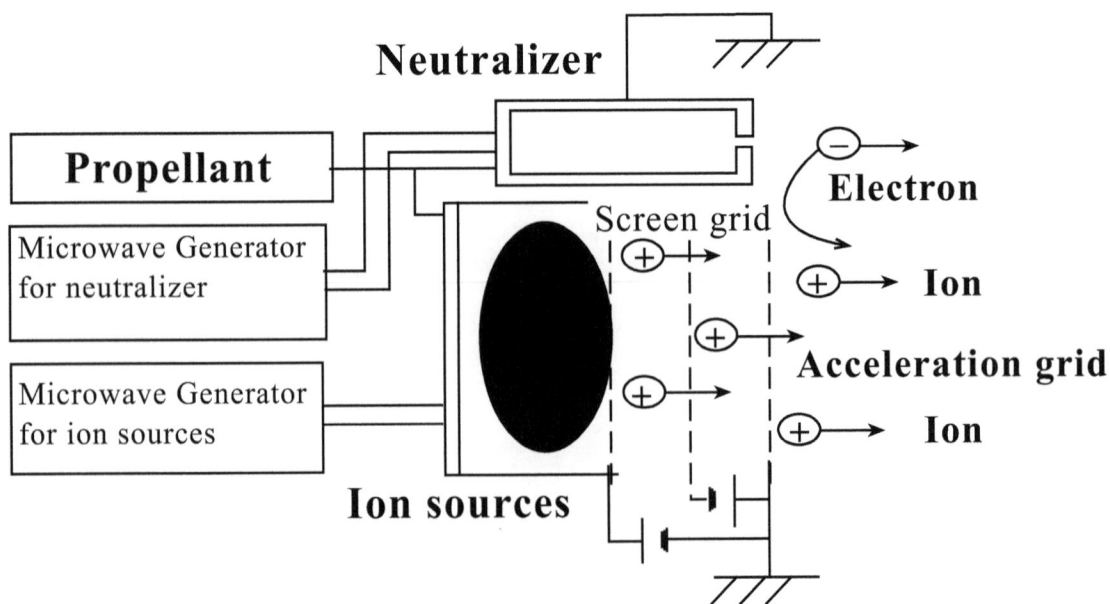

Figure 2: Microwave electric discharge ion thruster

Electrostatic acceleration is carried out by the potential difference of a screen grid and an accelerator grid as an ion beam. The ion generated by the 4.2GHz microwave supplied by the wave-guide tube from the microwave generator obtains a reactionary thrust.

The microwave electric discharge type ion thruster by the electron cyclotron resonance ECR can attain long-life compared with an electron impact type ion thruster since the electric discharge electrode is unnecessary and the number of power supplies can be reduced. This feature can make the thruster simple. This microwave electric discharge type ion thruster already has made "Hayabusa (MUSES-C)" to sail in space[1]. Hayabusa is a spacecraft exploring the asteroid Itokawa in the 'asteroid sample return' project conducted by JAXA (Japan Aerospace Exploration Agency) ISAS (Institute of Space and Astronautical Science) of Japan.

(c) Hall Thruster

A hall thruster has an annular acceleration channel, the electric field E is impressed in the direction of an axis and the annular magnetic field B is impressed radially (Fig. **3**).

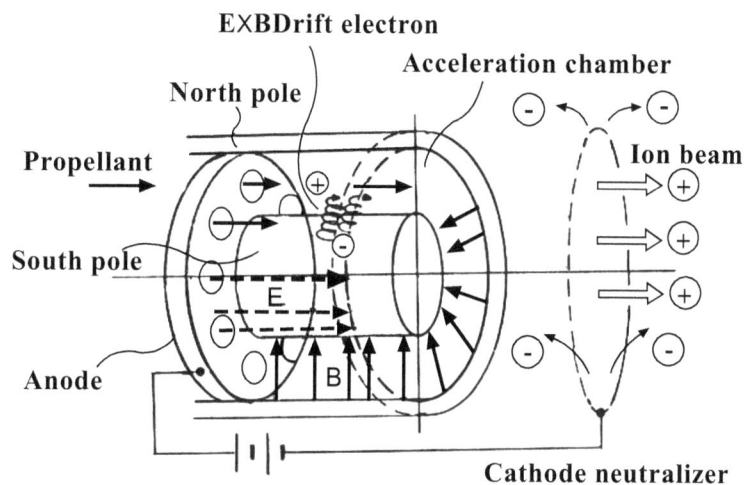

Figure 3: Hall thruster

The electron emitted from the negative pole performs what is called an $E \times B$ drift in the direction of a circumference within an acceleration channel by the interaction of the electric field E and the magnetic field B (hole current is induced) and generates ion by the ionization collision with a propellant. It is an electrostatic acceleration type engine which obtains thrust according to the reaction that the electrostatic acceleration of the ion is carried out by the electric field in the direction of an axis. On the other hand, since the electromagnetism acceleration by the Lorentz force of hall current and a magnetic field is also received, it can also be considered that hall thruster is combined electrostatic acceleration system and electromagnetism acceleration system. Since the inside of the acceleration channel is maintained at semi-neutrality by hall current and thrust restrictions of space charge limited current are not present as in the ion thruster, high thrust density can be attained.

A structure of hall thruster is comparatively lightweight and compact and there is an advantage of not needing the multiple high-voltage-direct-current power supplies like an ion thruster. Thrust is several 10 mN(s) to several 100 mN(s), a specific impulse is from 1000 seconds to 2000 seconds and it has propulsion efficiency in the order of 50%.

(d) MPD thruster

Electromagnetism acceleration plasma engine, *i.e.*, MPD (Magneto Plasma Dynamic) thruster has the negative pole (cathode) in the center and the coaxial electrode structure of the anode in the circumference. Propellant gas is ionized by arc electric discharge of large current due to the impressing high voltage pulse between the anode and the cathode. It is an engine which obtains a reaction thrust by carrying out acceleration injection of the generated plasma in back according to the Lorentz force of arc electric discharge current and magnetic field B_θ of the direction of a circumference induced by this current itself (Fig. **4**).

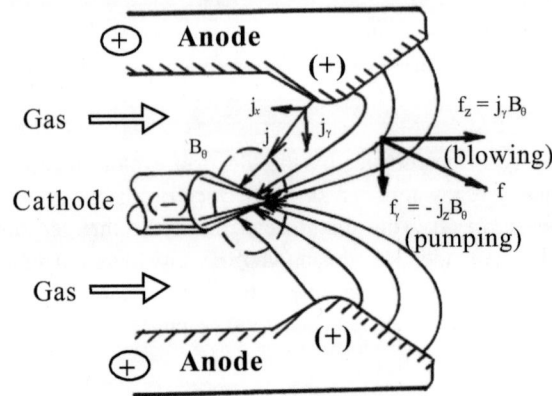

Figure 4: MPD thruster

The thrust of MPD thruster is composed of the blowing force in which the thrust acts in the direction of a direct axis and the pumping force which compresses plasma into a radius inner side, heightens the pressure at the tip of the negative pole and produces thrust as the reaction.

In order to generate Lorentz force effectively as electric discharge current, the large current more than the order of 1 kA is required. Its thrust density is comparatively large, a specific impulse is from 1000 seconds to 5000 seconds and the propulsion efficiency has the performance in the order of 20% to 40%.

In comparison to MPD thruster, arc jet thruster is an engine which has coaxial electrode structure like MPD thruster, carries out Joule heating of the propellant by arc discharge, carries out aerodynamic acceleration by a nozzle of gas at high temperature and high pressure and obtains thrust. Although arc discharge is used in both arc jet thruster and MPD thruster, acceleration of a propellant is aerodynamically performed in arc jet thruster and electromagnetically in MPD thruster.

Aerodynamic expansion acceleration will become dominant for the 1 kA or less of discharge current, vise versa, the electromagnetic-like acceleration will become dominant for the 1 kA or more of discharge current. By the way, the thrust of arc jet thruster is 50mN to several N, a specific impulse is from 500 seconds to 1000 seconds and the propulsion efficiency has the performance in the order of 30% to 40%.

ELECTRIC PROPULSION THEORY

Here, an outline is introduced about the performance of the common propulsion system containing electric propulsion. The foundations of the propulsion principle of the space propulsion system called a rocket are based on the momentum thrust (reaction thrust) induced by the law of conservation of momentum which injects operational substances, such as gas, back at high speed and is ahead propelled by the reaction.

The last attainment speed V (m/s) which can be reached when a rocket exhausts fuel (propellant) is given by the following equation, using a specific impulse Isp and acceleration due to gravity as g (s), the injection speed w of gas (m/s) and the mass ratio R:

$$V = w \cdot \ln R = g I_{SP} \cdot \ln R \,. \tag{2.1}$$

Usually, Eq. (2.1) is called Delta V (ΔV) and indicates the speed increment ΔV of a rocket. Mass ratio is the value which divide the total mass (initial mass) of the rocket at the time of fuel loading by all the rocket mass at the time of fuel full consumption (the last mass). Since R= 7 to 10 is a practical value and the natural logarithm of mass ratio R is $\ln R = 1.95$ *to* 2.3, it has a contribution to the rocket speed by mass ratio only about 1.95 to 2.3 times. The maximum speed which a rocket can reach is theoretically determined by the gas jet speed w (m/s) and its mass ratio R.

Modification of Eq. (2.1) is sometimes effective as follows:

$$R = \exp\left(\frac{V}{w}\right). \tag{2.2}$$

Specific impulse Isp(s) shows the value of the thrust F in which a propellant can generate per unit weight per second and is given by the following equation. $\dot{m}$ is the mass flux of a propellant in (kg/s).

$$I_{SP} = F / \dot{m} g = \dot{m} w / \dot{m} g = w / g \ (s). \tag{2.3}$$

Specific impulse is thrust per propellant unit flux, it is shown for how many seconds the thrust of a 1kg pile can be generated in a 1kg propellant and the unit is the second. The large specific impulse yields that as the gas jet speed w (m/s) becomes large the maximum speed of a rocket can be increased and fuel consumption becomes low. For this reason, a thruster with a high specific impulse becomes a planetary exploration spacecraft which cruises space for a long period of time.

From Eq. (2.3), thrust F (N) is given by

$$F = \dot{m} w = \dot{m} g I_{SP}. \tag{2.4}$$

where $\dot{m}$ (kg/s) is the mass flow rate of the propellant, w (m/s) is the exhaust speed.

The propulsion efficiency η is an important performance parameter and it shows how the injection power P was changed into propulsion power and is given by the following equation

$$\eta = \frac{(1/2)\dot{m} w^2}{P} = \frac{F^2}{2\dot{m}P}. \tag{2.5}$$

From Eqs.(2.3) and (2.5), Eq.(2.6) is obtained:

$$F = \frac{2\eta P}{g I_{SP}}. \tag{2.6}$$

Above-mentioned equation shows the performance common to a propulsion system driven by momentum thrust.

Below, a performance parameter peculiar to electric propulsion is shown. The performance index peculiar to electric propulsion are propellant utilization efficiency and ion production cost. Propellant utilization efficiency is a rate which shows how much the propellant flux supplied as a propellant is changed as an ion beam.

Propellant utilization efficiency is given by

$$\eta_u = \frac{\dot{m}_i}{\dot{m}_p}, \tag{2.7}$$

where $\dot{m}_i$ (kg/s) is ion beam flux, $\dot{m}_p$ (kg/s) is propellant supply flow rate.

Ion production cost is shown by the electric discharge electric power (W/A) which showed generation energy (eV/ion) required for per ion beam, or unit ion beam current generation took. Thrust F (N) is given using ion beam speed u_e (m/s) (it is the same as the exhaust speed w)

$$F = \dot{m}_i u_e. \tag{2.8}$$

A specific impulse *Isp*(s) is given by

$$I_{SP} = \frac{F}{\dot{m}_P g} = \frac{\dot{m}_i u_e}{\dot{m}_P g} = \left(\frac{\dot{m}_i}{\dot{m}_P}\right) \cdot \frac{u_e}{g} = \eta_u \frac{u_e}{g} .$$ (2.9)

In the case of electric propulsion, compared with Eq.(2.3), propellant utilization efficiency is applied as a coefficient. Since an exhaust operation substance is the same as a propellant in the case of a chemical rocket, η_u is set to 1 from Eq.(2.9), and it becomes Eq.(2.3).

Finally main performances, such as thrust of an ion thruster and MPD thruster, are shown as follows.

(a) Ion Thruster

Ion beam flux $\dot{m}_i$ (kg/s): $\dot{m}_i = \dfrac{M}{q} J_b$, (2.10)

Ion beam speed v_i(m/s): $v_i = \sqrt{\dfrac{2qV}{M}}$, (2.11)

Thrust F (N): $F = \dot{m}_i v_i = J_b \sqrt{\dfrac{2MV}{q}}$. (2.12)

Here, *M* is ion mass ($3.6 \times 10^{-27} kg$), and q is electric charge ($1.6 \times 10^{-19} C$) of ion and J_b (A) is beam current and *V* (v) is beam accelerating voltage.

(b) MPD Thruster

Thrust F (N): $F = \dfrac{\mu_0}{4\pi} J^2 \left(l_n \dfrac{r_a}{r_c} + \alpha \right) = k J^2$. (2.13)

Here, μ_0 is vacuum permeability, *J* is electric discharge current, r_a is anode radius, r_c is cathode radius, α is the value of 0 to 3/4 by the thrust coefficient. Thrust increases in proportion to the second power of electric discharge current J. The thrust coefficient $\alpha = 3/4$ shows the theoretical maximum thrust, and $\alpha = 0$ shows the theoretical minimum thrust.

LASER PROPULSION SYSTEM

Outline of Laser Propulsion

The basic principle of laser propulsion is the same as for a rocket except that acceleration of propellant is done by ablation using the laser irradiation. Any materials melt and evaporate when the irradiated by a high power laser radiation. Reaction thrust is generated due to vapor molecules or ions are ejected in the direction of pressure gradient formed on the material surface. This corresponds to the jet of the rocket and is a propulsion principle by the same momentum thrust as the rocket.

One of the most successful demonstrations of the practical use of laser propulsion was lead by L. Myrabo and F. B. Mead in November 1997. They succeeded in a vertical flight of 60 g mass to 15.25 m height with a maximum acceleration of 2.3g in an Air Breathing Pulse Detonation Laser method driven by a repetitive pulsed CO_2 laser of an average output power 10 kW [3,5].

The laser propulsion method is mainly classified into two kinds. The first one uses continuous wave (cw) laser for heating a propellant with a relatively high molecular weight such as water. Temperature of propellant stays relatively low and the efficiency of converting laser energy to momentum is high. The second one typically uses pulsed laser radiation to generate high temperature vapor or even ionized gas of light molecular weight such as hydrogen. One of the advantages of laser propulsion is that driving energy of propellant can be delivered from a remote laser station and no energy source needs to be on-board resulting in an increased payload. This method might be acceptable for the rocket launch from the ground or moon mission where a laser beam from a fixed laser station can reach the space craft while the beam diameter remains in reasonable size for a laser power receiver before beam divergence becomes larger than space craft. For example, if one assumes the diameters of laser transmitter and receiver to be 10 and 5 meter respectively for diffraction limited infrared laser beam, the maximum transmitting distance would be 150,000 km. Beyond this distance, power receiving efficiency decreases to the square of the transmitting distance. Therefore, on-board power supply and laser system become a reasonable choice for any mission within solar system and beyond unless many laser power stations are distributed along space craft orbits which does not seem feasible. Besides beam divergence, keeping beam tracking between very long distance in free space is a challenging engineering task. It is desirable that a free flight with a compact energy source for laser driver becomes possible. This is already possible considering the recent developments of high power laser diodes and the miniaturization of a power supply. The electric power of a practical ion engine has been already supplied by the power source equipped on a satellite and spacecraft. More than an equivalent amount of thrust to ion engines can be generated by laser system connected with the current on-board power supply system.

An example of realizing such pulsed laser is a Diode Pumped Solid State Laser, DPSSL as shown in Fig. **5** where LDs pumps a Q-switched solid state laser. The solid laser can easily generate the pulse of light by the combination of Q switch element and crystal laser medium such as Nd:YAG that can accumulate large energy and emits radiation energy instantaneously. The block diagram of the laser system is shown in Fig. **6.**

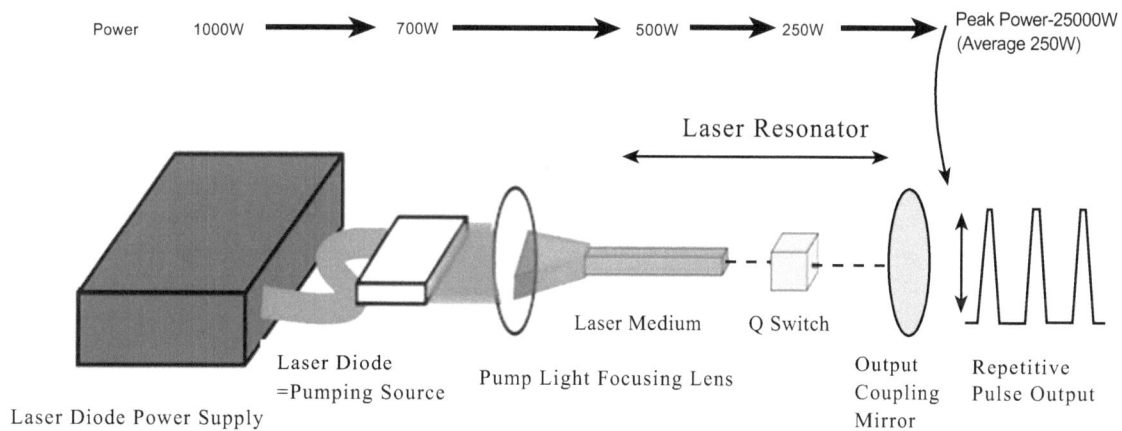

Figure 5: Power pulse laser generation device

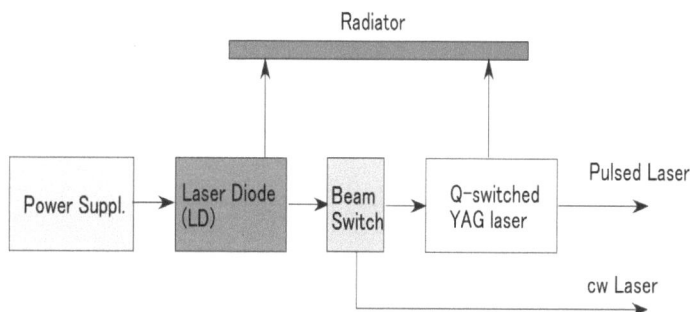

Figure 6: Block diagram of the laser system

Laser Propulsion Theory

In laser propulsion technology, velocity and fluid conditions of propellant can be controlled by the energy density of laser radiation by means of the combination of laser parameters such as wavelength and intensity and propellant.

There are two important performance parameters defined for the thrust generation by laser propulsion scheme. They are the specific impulse I_{SP} used in the conventional propulsion scheme and momentum coupling coefficient C_m $(N \cdot s/J = N/W)$ specific to laser propulsion [2,7].

$$C_m = \frac{F}{P_L} = \frac{\dot{m}V_E}{P_L},$$

(3.1)

$$I_{SP} = \frac{F}{g\dot{m}} = \frac{\dot{m}V_E}{g\dot{m}} = \frac{V_E}{g}.$$

(3.2)

where C_m is defined as a ratio of thrust F to incident laser power P_L. It is similar to the thrust electric power ratio (N/kW) of the electric propulsion.

Further, similarly to the electric propulsion, the following relation between thrust F and propulsion efficiency η, which is defined as the ratio of propellant kinetic energy to the laser energy, is important.

$$F = \frac{2\eta P_L}{g I_{SP}},$$

(3.3)

where g is gravitational acceleration and P_L is laser power.

In addition to these, a fraction of laser power, P_L converted to propellant kinetic power; $\dot{m}V_E^2 / 2$ is defined,

$$\eta = \frac{\frac{1}{2}\dot{m}V_E^2}{P_L}.$$

(3.4)

Combining Eqs.(3.1),(3.2) and (3.4) gives the relation among these parameters;

$$C_m \cdot I_{SP} = \frac{F^2}{P_L g\dot{m}} = \frac{\dot{m}^2 V_E^2}{P_L g\dot{m}} = \frac{2}{g}\frac{\frac{1}{2}\dot{m}V_E^2}{P_L} = \frac{2}{g}\eta$$

(3.5)

Equation (3.5) shows the fluid efficiency. The performance of laser propulsion is controlled by η as an upper bound. On the other hand, specific impulse I_{SP} is determined only at speed V_E of the propellant from Eq.(3.2). Because C_m and I_{SP} are inversely proportional to each other from the Eq.(3.5), when fluid efficiency $(2\eta/g)$ is determined, a trade-off of C_m and I_{SP} becomes possible. Therefore, a trade-off between efficiencies of the laser power and the propellant is possible. For instance, it is shown that the propulsion system that produces thrust by a little amount of propellant, *i.e.*, large specific impulse (large I_{SP}) and produces large thrust to the laser power (large C_m because a large amount of propellant is necessary) can design laser propulsion while reflecting the condition of reality.

In other words, selection between high C_m system and high I_{SP} system can be realized by controlling propellant exhausting velocity. The maximum energy efficiency and performance, "energy saving mode" or "propellant saving mode" of a laser thruster are determined by η and the exhaust velocity of propellant. The performance of laser propulsion scheme is not limited by the "specific" chemical energy of fuel. Thermal energy of propellant can be controlled by the ratio between propellant mass flow rate and laser power both of which are easily controlled without installing any complex devices. The propellant saving mode of high I_{SP} is the best choice for the transition to the geosynchronous orbit from a circular orbit or a deep space mission.

In the case of pulsed laser mode, since one pulse energy E (J), pulse duration $\tau(s)$, pulse repetition rate f (Hz), average laser power P_{ave} and peak laser power P_{pk} have the following relations

$$E = P_{pk}\tau \ , \qquad P_{ave} = E \ f \ .$$
(3.6)

Substituting average laser power P_{ave} for P_L in Eq.(3.3), thrust F is given by

$$F = \frac{2\eta P_{pk}\tau \ f}{gI_{SP}} \ .$$
(3.7)

One of the authors (Minami) and others investigated a laser propulsion vehicle that irradiated the laser to ice or water and generated thrust from the steam explosion [4,8].

NUCLEAR PROPULSION SYSTEM

A nuclear propulsion system is classified into the heat nuclear fission propulsion system, nuclear fusion propulsion system, and anti-matter propulsion system which give nuclear heat to a propellant and obtain thrust. The nuclear propulsion system has a reactor core in the portion equivalent to the combustion chamber of chemical propulsion system, and propels using the nuclear energy generated by a nuclear reaction.

Hydrogen is used as a propellant. The temperature and a specific impulse are determined by the energy which per the unit mass is heated. The maximum of temperature is decided from the safety of a reactor core or its maintenance system, and they are about 3000K. The specific impulses obtained by nuclear propulsion are 800 seconds to 950 seconds and it is greater about 2 times of liquid oxygen and liquid hydrogen engine (450 seconds). The pressure of the thrust chamber is almost comparable as chemical propulsion, after it pressurizes propellant with turbo pump, cools the wall and nozzle wall of a thrust chamber and is guided to a thrust chamber. Since the diameter of a nozzle is also comparable as chemical propulsion, a generating thrust is set to a chemical propulsion level. Since it has a reactor core in a thrust chamber and has the cover material of radiation, a thrust weight ratio is comparatively small. The nuclear propulsion system has large acceleration and high speed; therefore, it fits a manned mission which has restriction of flight time. An outline is introduced below.

Nuclear Fission Propulsion System

The nuclear fission propulsion system which propels by carrying out heating expansion injection of the propellant with nuclear fission reaction heat is classified into a solid core type, a particle bed type, a liquid core type, and a gas core type according to the form of fuel at the heat.

As an example, a liquid core type propulsion engine is shown in Fig. 7. Liquid fuel temperature in engine is made high more than the melting point of a little less than 300-degree uranium compound. Liquid fuel is held with centrifugal force in a rotation drum, injection pouring of the hydrogen is carried out there, and hydrogen is heated there. A specific impulse is from 1300 seconds to 1500 seconds and thrust weight ratio is one or more.

LIQUID-CORE NUCLEAR ROCKET

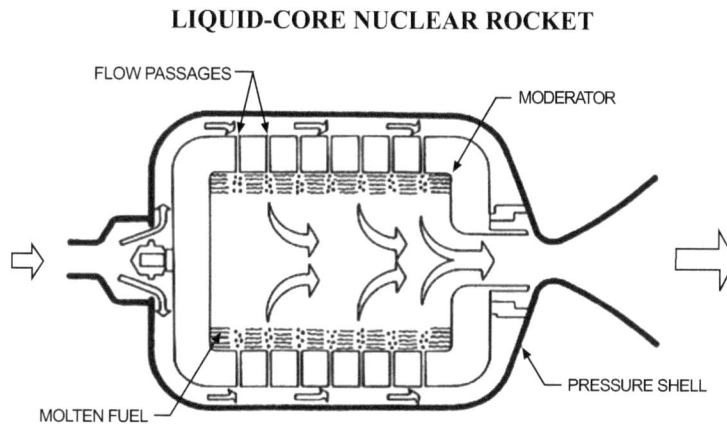

Figure 7: Liquid core type propulsion engine

Nuclear Fusion Propulsion System

The two following propulsion systems which propels by using the reaction heat of nuclear fusion and carrying out heating expansion injection of the propellant are examined.

(a) Steady-State Fusion Reaction System

A hydrogen propellant is made to mix in the plasma fuel by which magnetic maintenance was carried out in a steady-state fusion reaction system. A specific impulse is said to be 2,500 second to 200,000 seconds and thrust weight ratio is called 10^{-5} to 10^{-4}. Although various methods are proposed by the magnetic plasma confinement system, as fuel, D-^{3}He fuel with little neutron generating is used. Recently, FRC (Field-Reversed Configuration) is carried out. The concept of a FRC rocket is shown in Fig. **8**.

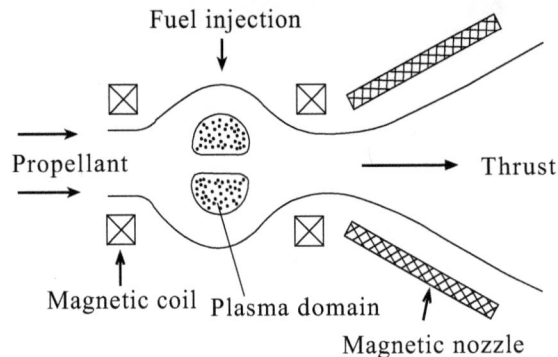

Figure 8: FRC fusion rocket

(b) Inertia Fusion Reaction System

The nuclear fusion which irradiates powerful laser or a heavy ion beam to a small (D-T) fuel pellet is called pulse nuclear fusion or inertia nuclear fusion and can be considerd a propulsion system by this method.

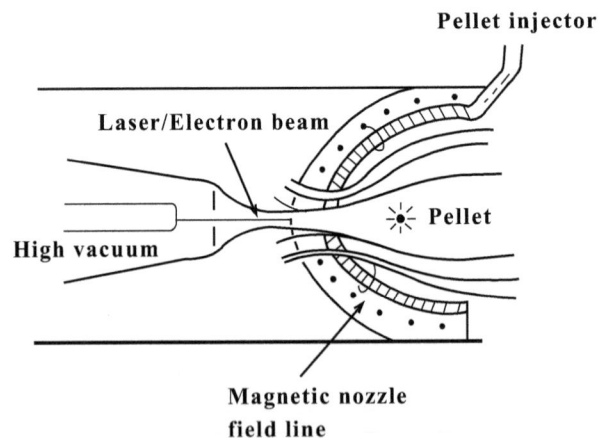

Figure 9: Fusion pulse rocket

The concept is shown in Fig. **9**. There exists a high specific impulse type and a low specific impulse type in inertia nuclear fusion. The performance is 100,000 MW of outputs, thrust 970kN and specific impulse 6,320 seconds, engine weight less than 200ton, the thrust weight ratio is 0.49.

ANTI-MATTER PROPULSION SYSTEM

An anti-matter propulsion system is a rocket which generates the thrust using the huge reaction thermal energy by the annihilation reaction of a particle (a proton, hydrogen) and an antiparticle (an antiproton, anti-hydrogen). Various systems are proposed.

The anti-matter propulsion system is classified into a liquid core type rocket, a solid core type rocket, a gas core type rocket, a plasma core type rocket, a pie-on (Pion) rocket and a Photon rocket according to the physical condition of propellant. The systems which can reach the quasi-speed of light near the velocity of light in this are only a Pion rocket and a Photon rocket. However, since an acceleration performance is as extremely low in the order of 10^{-3}G to 10^{-2}G and is a low thrust, it will continue accelerating for a long period of time for tens of years. The concept of a solid core type rocket is shown in Fig. **10**, and the concept of a plasma core type rocket is shown in Fig. **11**.

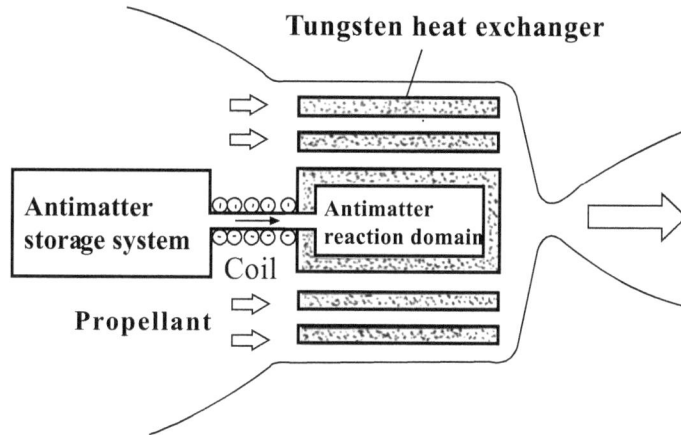

Figure 10: Antimatter core rocket

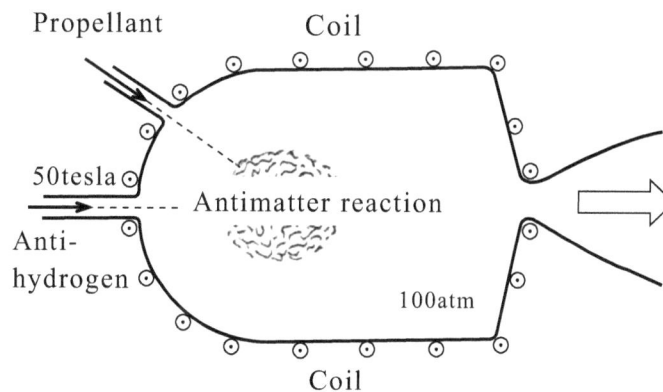

Figure 11: Antimatter plasma core rocket

A solid core type rocket propels by carrying out heating expansion injection of the propellant which passes the wall of tungsten using the huge reaction heat induced by the annihilation reaction of matter and anti-matter. The plasma core type rocket propels by injection of the charged pi-meson (Pion) generated by the annihilation reaction of anti-hydrogen and the propellant according to the magnetic mirror effect.

The conceptual figure of a Pion rocket is shown in Fig. **12**. The outline of the propulsion principle using Pion is as follows. The first step of annihilation reaction for anti-proton and proton is a Pion generation.

$$\bar{P} + P \rightarrow \pi^0 + \pi^+ + \pi^-$$

Although the life-time of π^0 is as short as 8.3×10^{-17} seconds, the life-time of π^+, π^- as a charged pi meson is far as 2.6×10^{-8} long. So, if there exists a magnetic slope structure in a charged pi meson generation domain, the charged pi meson such as π^+, π^- will be injected to the domain where magnetism is small (application of the magnetic mirror effect). As shown in Fig. **12**, an antiproton is poured from engine left-hand side and hydrogen ion gas (proton P) is irradiated from a lower part. An injection of charged pi meson (π^+, π^-) is carried out in accordance with a magnetic slope as above-mentioned in an annihilation domain and a reaction thrust is obtained. The mass of Pion is about 1/7 of a proton.

A Photon rocket is a system which obtains a thrust by light pressure of photon (γ quantum) produced in the last reaction course of the annihilation reaction of a matter and an anti-matter to the reflector. The conceptual figure of a Photon rocket is shown in Fig. **13**. The specific impulse in a Pion rocket and a Photon rocket is high and the maximum speed of a rocket is theoretically possible to the quasi- speed of light, however, since an acceleration performance is as low as about 10^{-3} G to 10^{-2} G and is a low thrust, it will continue accelerating for a long time (the acceleration time for 9.7 years is required for 10% attainment of the speed of light at 10^{-2} G).

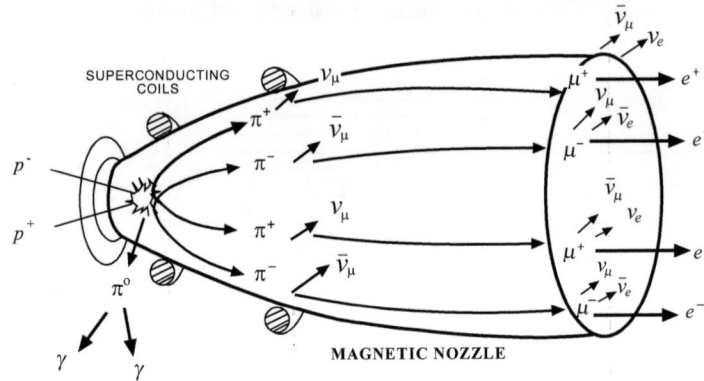

Figure 12: Pion rocket

As above-mentioned, the energy of 100 times or more of nuclear fusion energy can be taken out from an antiproton and a proton annihilation reaction. Further, a miniaturization of engine is possible and radiation of neutron is not given off, therefore, the reactor of a safe side becomes comparatively possible. As a subject of technical development for this, technology which generates large amount of antiproton and storing technology of antiproton as a fuel source for a long period of time must be put in practical use.

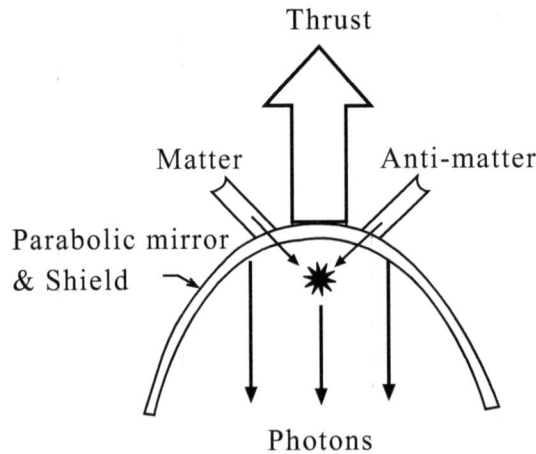

Figure 13: Photon rocket

TECHNICAL STATUS AND PROBLEM FOR PRESENT PROPULSION SYSTEM

As described above, all kinds of current propulsion system except solar sail are based on the momentum conservation law. In the case of the momentum thrust based on momentum conservation law, the maximum speed (V) is limited by the product of the gas effective exhaust speed (w) and the natural logarithm of mass ratio (R)

$$V = w \cdot \ln R = g I_{SP} \cdot \ln R .$$

                                                   (5.1)

The maximum speed V which a rocket can reach is theoretically determined by the gas jet speed w (m/s) and the mass ratio R.

Because the velocity of present rocket is too slow as compared with the speed of planets, the interplanetary exploration by Mankind, not to speak of interstellar exploration, has various technical difficulties. We need the super-high speed and high acceleration of spaceship.

For example, the origin of the problem that the manned Mars exploration takes long-term time is due to the cruise velocity of a spacecraft is too slow. The second astronomical speed (11.2 km/s) that a rocket obtains for earth escape is slightly slow compared with the orbital speed (24 km/s) of Mars, and the orbital speed (30 km/s) of the Earth.

This is because the maximum speed of a rocket is limited by the product of gas effective exhaust speed and the natural logarithm of mass ratio (its value is about 7). The speed beyond this limit cannot be theoretically taken out from the propulsion principle of a rocket based on the momentum conservation law.

Concerning a chemical rocket which has multi-stage composition, about 10km/s speed is a practical limit. In the case of chemical rocket, its I_{SP} is 460 seconds, so the maximum speed becomes 4.5km/s for single stage rocket. If the speed of a rocket is 1000 times quick compared with the speed of Mars or the Earth, a straight line orbit can be attained. Whenever you wish, it can always reach to the target planet in a short time without restriction of orbital calculation, a start time and return time, so that it may operate as if it were a car.

Equation (5.1) can be represented as follows in detail:

$$V_f - V_i = \Delta V = \int_0^T \alpha dt = \int_0^T \frac{F}{m} dt = \int_0^T \frac{I_{SP}(-\dot{m}g)}{m} dt = I_{SP}g \ln \frac{m_i}{m_f}. \tag{5.2}$$

Eq.(5.2) indicates that the speed increment ΔV of a rocket when its initial mass m_i reduces to its the final mass m_f by combustion for T seconds. Since the propellant mass m_p is given by Eq.(5.3), combining Eqs.(5.2) and (5.3) yields the Eq.(5.4).

$$m_p = m_i - m_f, \tag{5.3}$$

$$m_p = m_i \left[1 - \exp\left(-\frac{\Delta V}{g I_{SP}} \right) \right]. \tag{5.4}$$

By expelling the mass of a propellant m_p outside, a rocket obtains and promotes speed ΔV, that is, the propellant is indispensable for the conventional propulsion system based on momentum thrust. Further, since a large thrust is required for a heavy payload, a large amount of propellant is needed for rocket; therefore the increased weight of propellant, *i.e.* the rocket needs larger amount of propellant.

Accordingly, we need the new propulsion principle to exceed the limits of prior propulsion technology and seek entirely different technology.

As described in NASA BPP, we must investigate the new propulsion system based on physics as follows.

1. **Mass:** Discover new propulsion methods that eliminate the need for propellant or beamed energy.

2. **Speed:** Discover how to circumvent existing limits (light-speed) to dramatically reduce transit times.

3. **Energy:** Discover new energy methods to power these propulsion systems.

CHAPTER 3

Advanced Physics for Field Propulsion

Abstract: Field propulsion is the concept of propulsion theory of spaceship that is not based on usual momentum thrust but based on momentum derived from an interaction of the spaceship with external fields. Field propulsion system is propelled without mass expulsion. The propulsion principle is based on the assumption that space as a vacuum possesses a substantial physical structure. Field propulsion is propelled receiving the propulsive force (*i.e.*, thrust) arises from the interaction of the substantial physical structure. The meaning of substantial physical structure regarding the space-time can be conjectured from both General Relativity in the view of macroscopic structure and Quantum Field Theory in the view of microscopic structure. Therefore, several kinds of field propulsion can be proposed by making choice of physical concepts.

In this chapter, the basic concepts regarding the structure of physical space for such means of propulsion within the present physics paradigm are described.

INTRODUCTION

Field propulsion is the concept of propulsion theory of spaceship not based on usual momentum thrust but based on momentum derived from an interaction of the spaceship with external fields. The field propulsion principle is based on the assumption that space as a vacuum possesses a substantial physical structure. Field propulsion is propelled receiving the propulsive force (*i.e.*, thrust) arises from the interaction of the substantial physical structure.

Instead of conventional chemical propulsion systems, they are based on the General Relativity Theory, Quantum Field Theory and the other exotic theory. In this chapter, the basic concepts for such means of propulsion within the present physics paradigm are presented.

PROPULSION PRINCIPLE OF FIELD PROPULSION

Background

As has been previously described in chapter 2, all existing methods of propulsion systems, *i.e.* chemical propulsion, electric propulsion (Ion thruster, MPD [Magneto Plasma Dynamic] thruster, Hall thruster, ARC jet thruster), laser propulsion, nuclear propulsion are based on expulsion of a mass to induce a reaction thrust. The "momentum thrust" is based on momentum conservation law [15].

Alternatively, the concept of "Field Propulsion" has been advanced as resulting from pressure thrust, without mass expulsion. The envisaged solar sails and light sails are propelled just by receiving light pressure, but pressure thrust in Field Propulsion refers to a reaction with space-time itself (*i.e.* the vacuum) to generate a propulsive force. The propulsive force as a pressure thrust arises from the interaction of space-time around the spaceship and the spaceship itself, the latter being propelled against space-time structure. The Field Propulsion principle consists in the exploitation of the action of the medium field induced by such interaction and is thus based on some concepts in modern physics to be found in General Relativity, Quantum Field Theory, Quantum Cosmology and Superstring Theory including D-brane in order to bring about the best propulsive performance.

The most remarkable results attainable through Field Propulsion are as follows: 1) high acceleration such as several ten G can be obtained, 2) theoretical final velocity close to the speed of light, 3) no action of inertial force.

As to item 3), this comes from the thrust as a body force. Since the body force they produce acts uniformly on every atom inside the spaceship, accelerations of any magnitude can be produced with no strain on the crew, *i.e.*, it is equivalent to free-fall. Therefore, the flight patterns such as quick start from stationary state to all directions in the atmosphere, quick stop, perpendicular turn, and zigzag turn are possible.

For instance, as an example of field propulsion system, the Space Strain Propulsion System [9] and alternatively the Space Drive Propulsion System [10] derived from General Relativity were firstly introduced by Y. Minami. The

Takaaki Musha and Yoshinari Minami

expression "space strain" was changed to "space drive" after recommendation by R.L. Forward. Further, Space Coupling Propulsion System was introduced by Marc G. Millis [12]. At present, physics admits the Zero-Point Energy (Media of Electromagnetic Fluctuations of the Vacuum: Zero-Point Fluctuations or Zero-Point Field). It is said that the Zero-Point Field is related to both gravitation and inertia. Space as a vacuum is a kind of actual field, which repeats the creation and annihilation of particle and anti-particle continuously. ZPF (Zero-Point Field) propulsion system was introduced by H.D. Froning and T.W. Barrett [3,4]. The various standpoints for engineering of the Zero-Point Field were introduced by H.E. Puthoff [17]. Also, a highly desirable feature for interstellar travel induced by the distortion of the local space-time metric in the region of a spaceship was shown by M.Alcubierre [1].

The field propulsion system appears to violate the conservation law of momentum because the reaction mass is not readily apparent. NASA considers that conservation of momentum can be satisfied in various ways that do not require having an on-board supply of reaction mass about conservation of momentum, as follows [12]:

1) conservation by using the contents of space (Interstellar Matter, Star Light, Magnetic Field, Cosmic Microwave etc.) as the reaction mass, 2) conservation by expelling non-mass momentum (equivalent momentum $P=E/c$) such as hypothetical "Space Waves", 3) conservation by negative mass, 4) conservation by coupling to distant masses via the intervening space.

However, we propose that the most promising interpretation is to consider that space itself as vacuum is a kind of reaction mass.

Now, a field propulsion system must satisfy the following criteria [13]:

1) conservation of momentum, 2) conservation of energy, 3) ability to induce a unidirectional acceleration of the spaceship, 4) controllability of direction and thrust, 5) sustainability during spaceship motion, 6) effective capability of propelling the spaceship.

Generic Propulsion Principle of Field Propulsion

As shown in Fig. **1**, the propulsion principle of field propulsion system is not momentum thrust but pressure thrust induced by pressure gradient (or potential gradient) of space-time field (or vacuum field) between bow and stern of spaceship. Since the pressure of vacuum field in the rear vicinity of spaceship is high, spaceship is pushed from vacuum field. Pressure of vacuum field in the front vicinity of spaceship is low, so spaceship is pulled from vacuum field. In the front vicinity of spaceship, the pressure of vacuum field is not necessarily low but the ordinary vacuum field, that is, just only a high pressure of vacuum field in the rear vicinity of spaceship. The spaceship is propelled by this distribution of pressure of vacuum field. Vice versa, it is the same principle that the pressure of vacuum field in the front vicinity of spaceship is just only low and the pressure of vacuum field in the rear vicinity of spaceship is ordinary. In any case, the pressure gradient of vacuum field (potential gradient) is formed over the entire range of spaceship, so that the spaceship is propelled by pushing from the pressure gradient of vacuum field.

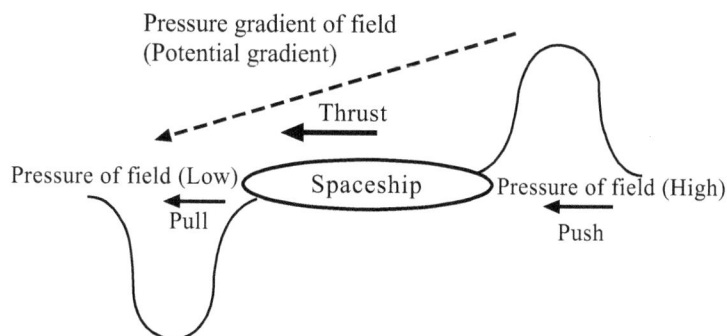

Figure 1: Fundamental propulsion principle of Field propulsion

Here, we must pay attention to the following. Spaceship cannot move unless the spaceship is independent of pressure gradient of vacuum field. No interaction is present between pressure gradient of vacuum field and spaceship. Spaceship does not move as long as the propulsion engine generates the pressure gradient or potential gradient in the surrounding area of spaceship, due to the interaction between pressure gradient of vacuum field and spaceship. This is because an action of propulsion engine on space is in equilibrium with a reaction from space. It is consequently necessary to shut off the equilibrium state in order to actually move the spaceship. As a continuum, the space has a finite strain rate, *i.e.* speed of light. When the propulsion engine stops generating the pressure gradient of vacuum field, it takes a finite interval of time for the generated pressure gradient of vacuum field to return to ordinary vacuum field. In the meantime, the spaceship is independent of pressure gradient of vacuum field. It is therefore possible for the spaceship to proceed ahead receiving the action from the vacuum field.

In general, a body cannot move carrying, or together with, a field that is generated by its body from the standpoint of kinematics. In other words, the body cannot move unless the body is independent of the field. This is because an action on the field and a reaction from the field are in the state of equilibrium.

As mentioned above, since the propulsion engine must necessarily be shut off for propulsion, the spaceship can get continuous thrust by repeating the alternate ON/OFF change in the engine operation at a high frequency.

Concerning the propulsion principle of field propulsion system, the distribution of field as shown in Fig. **1** is fundamental; accordingly, several kinds of propulsion systems have been proposed. Even if any propulsion system is selected, whether the constituents of pressure gradient or potential gradient generated by propulsion engine are curvature, metric, zero-point radiation pressure or entropy, the propulsion principle of field propulsion system is the identical. Y. Minami summarized the basic concept of field propulsion in 2003 [11].

Further, as is already explained in Section 2.5, all propulsion system based on the momentum thrust receives the reaction thrust by expelling the propellant mass. However, since no propellant is necessary for field propulsion, field propulsion is well called a propellant-less propulsion.

BASIC PHYSICAL CONCEPTS APPLIED FOR FIELD PROPULSION

As has been previously mentioned in 3.1, the propulsion principle of field propulsion is based on the assumption that space as a vacuum possesses a substantial physical structure. Field propulsion is propelled receiving the propulsive force (*i.e.*, thrust) arises from the interaction of the substantial physical structure. Then, what is the meaning of substantial physical structure regarding the space-time? The answer can be conjectured from both General Relativity in the view of macroscopic structure and Quantum Field Theory in the view of microscopic structure as shown in Fig. **2**.

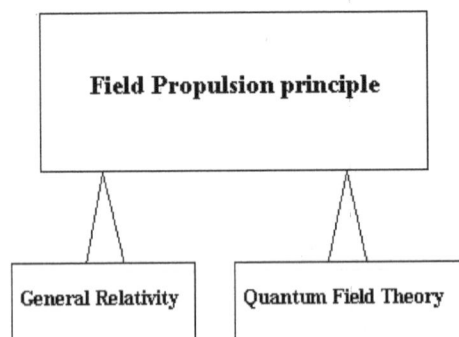

Figure 2: Relation between Physical concepts and Field propulsion

Firstly, General Relativity is the geometric theory of gravitation and the gravitation is explained by curved space. The curvature of space plays an important role. Although, the curvature is quantity in mathematics, the curvature has relations with continuum mechanics such as expansion, contraction, elongation, torsion and bending. This physical relation indicates that the space as a vacuum can be considered as a kind of elastic body like rubber in continuum mechanics. Therefore, the propulsion system used General Relativity is to be proposed from the standpoint of continuum mechanics.

Secondly, space-time as a vacuum is generally viewed as a transparent and ubiquitous infinitive empty continuum, upon which physical events take place. However, quantum field theory and quantum electrodynamics (QED) views it as possessing vigor and vitality over scales of time and space. Such vigor and vitality are the zero-point fluctuations of the vacuum electromagnetic field (vacuum perturbation), and the continuous creation and annihilation of virtual particle pairs. Further, according to the latest quantum optics, although until recently, it has been considered that the control of vacuum perturbation was utterly impossible, at present, it is proven that the vacuum perturbation can be controlled by squeezed light technology. At the present day, it is possible to increase the energy density locally above the vacuum state and vice versa, decrease the energy density locally below the vacuum state. That is, the squeezed light generates the squeezed vacuum states and yields the coordination geometry of energy density.

Furthermore, the strings of superstring theory are considered as the threads of the space-time fabric. String seems to be the fundamental element of the substructure or fine structure of space-time. Supposing that the string is the constituent of space-time is suggestive of the existence of possible quantum states for space-time. This indicates that entropy of space-time can be defined as an assembling of strings. Strings as the constituents of space-time correspond to the polymer chains in the elastic body. Since the statistical entropy is the logarithm of the number of states (*i.e.*, degeneracy of system), it is necessary to consider what kinds of physical state exist.

Therefore, the propulsion system used Quantum Field Theory is to be proposed from the standpoint of quantum physics.

As stated above, several kinds of field propulsion can be proposed by making choice of physical concepts. However, any propulsion system is selected, whether the constituents of physical structure are curvature, zero-point fluctuations, or statistical entropy of string and so on, the propulsion principle of field propulsion system is the identical in regard to utilize the substantial physical structure of space.

OUTLINE OF GENERAL RELATIVITY

In this section, we run through the nutshell of General Relativity to understand the field propulsion principle in brief.

General Relativity is the geometric theory of gravitation formulated by Albert Einstein in 1916, which was an extension of Special Relativity. It unifies special relativity and Newton's law of universal gravitation, and describes gravity as a geometric property of space and time, or space-time. In particular, the curvature of space-time is directly related to the four-momentum (mass-energy and linear momentum) of whatever matter and radiation are present. The relation is specified by the Einstein field equations, a system of partial differential equations. Many predictions of general relativity differ significantly from those of classical physics, especially concerning the passage of time, the geometry of space, the motion of bodies in free fall, and the propagation of light. Einstein's theory has important astrophysical implications, that is, general relativity is the basis of current cosmological models of a consistently expanding universe.

What is gravity? The answer is that the gravity is explained by curved space-time. From the standpoint of mathematical method, Riemannian geometry and tensor analysis are required. Especially, the metric tensor is important and fundamental. The notion of tensor is extended from the notion of vector. The brief explanation is described below.

The quantities $T_{iklm\ldots}{}^{rst\ldots}$ in which the indices can, independently, take on the values 1,2,3,4 are called tensor components. They are, in particular, called covariant components for the indices *iklm...* and contravariant for the indices *rst...* .

The number of indices which the components have is called the rank of the tensor. Tensors of first rank are also called vectors. The example of vector is the coordinates x^i of a point (x^1, x^2, x^3, x^4 for 4 dimensional spaces). A vector is a tensor whose components are equal in number to the dimensionality of the space in which it is defined. Tensors are mathematical objects that represent a generalization of the vector concept. As well, a tensor of zero rank is called scalar.

A tensor of second rank a_{ik} and a vector x^k can be combined to give the vector $z_i = a_{ik}x^k$. There are many different types of tensors, such as the strain tensor describes the local distortion of material, metric tensor, and curvature tensor, which has twenty components, describes the deviation of space-time continuum from flatness.

Since the components of vector is three in three dimensional space (F_i: i=1,2,3), the components of tensor of second rank is enlarged to nine, as (F_{ij}: i=1,2,3, j=1,2,3). F_{ij} is called second rank tensor. As might be expected, in the case of tensor, there are third rank tensor, … n-th rank tensor. The metric tensor is usually expressed g_{ij} .

The metric tensor of Special Relativity which shows flat space-time is a constant, and takes constant value regardless of a place. However, metric tensor of the curved space is a function of the place which changes a value by a place.

In General Relativity, mass particles move the geodesic line of space-time. A geodesic line is a curve which connects two on a curved surface with the shortest distance. Although the trajectory of a particle serves as a straight line in flat space, the trajectory of a particle serves as a curve in the curved space-time with gravity. That is, a particle moves in accordance with the following geodesic equation:

$$\frac{d^2 x^\mu}{d\tau^2} + \Gamma^\mu{}_{\nu\lambda} \frac{dx^\nu}{d\tau} \frac{dx^\lambda}{d\tau} = 0 \tag{3.1}$$

In the curved space, a result from which the trajectory differed is brought by which course even if particles move a closed curve, is chosen. Even if it goes around, it cannot return to the starting point of a basis.

For example, in the space-time at which it turned in on the earth, even if it carries out parallel displacement of the vector on between space-time, the results of parallel displacement differ by along which route it passed. In order to explain this phenomenon, curvature tensor (Riemann tensor) plays a role.

A coordinate system serves as a curvilinear coordinate system fundamentally. Thus, it is the greatest feature of the General Relativity to have the curvature. The Riemannian geometry which shows the geometrical structure of such curved space-time mathematically serves as an important mathematical means. The physical space in which we live in is the Riemann space-time.

Then, what this curvature does to a reason? If energies, such as mass energy of a mass object, electromagnetic energy, and thermal energy, exist, space-time will be curved. It is called a gravitational field equation alias Einstein equation to express this relation.

$$R^{ij} - \frac{1}{2} \cdot g^{ij} R = -\frac{8\pi G}{c^4} \cdot T^{ij} \tag{3.2}$$

where R^{ij} is the Ricci tensor, R is the scalar curvature, g^{ij} is metric tensor, G is the Newton's constant, c is the speed of light, and T^{ij} is the energy momentum tensor.

The left side shows the bend condition of space-time and the right-hand side shows the energy source to generate the curvature of space-time.

The basic preconditions of the General Relativity are the general principle of relativity and an equivalence principle. Firstly, all the organic law of physics is equally materialized in all the coordinate systems, *i.e.*, the general principle of relativity is that a physical law is invariance to general coordinate conversion. Secondary, an equivalence principle is saying that the size of gravity is changeable with suitable coordinate conversion, and the true gravity by the earth and the apparent force of the force of inertia by coordinate conversion cannot be identified.

The general laws of physics can be expressed in a form which is independent of the choice of space-time coordinates. For this reason, the tensor analysis is effective.

As described above, the flat space-time of Minkowski for Special Relativity is to be replaced by a curved space-time, and the curvature is to be responsible for gravitational effect in General Relativity.

The difference can be explained as whether space is curved or not, that is, whether 20 independent components of

Riemann curvature tensor are zero or not. After all, the existence of space curvature determines whether the object drops straight down or not. Although the space curvature at the surface of the Earth is very small, *i.e.*, $3.42 \times 10^{-23} (1/m^2)$, it is enough value to produce 1G (9.8m/s^2) acceleration. On the contrary, the space curvature in the universe is zero; therefore any acceleration is not produced. Accordingly, if the space curvature of a localized area including object is controlled to $3.42 \times 10^{-23} (1/m^2)$ curvature, the object moves receiving 1G acceleration in the universe. The square of the infinitesimal distance "*ds*" between two infinitely proximate points x^i and $x^i + dx^i$ is given by equation of the form

$$ds^2 = g_{ij} dx^i dx^j$$

(3.3)

where g_{ij} is metric tensor.

The metric tensor g_{ij} is a quantity defined geometrical character of space and is functions of x^i in general. The metric tensor is constant in the case of flat space (*i.e.* Special Relativity represented by the Minkowski metric η_{ij}). In a curved Riemannian space, the property of space depends on the metric tensor g_{ij}. The infinitesimal distance ds^2 is given by metric tensor g_{ij}, and also metric tensor g_{ij} determines Riemann connection coefficient Γ^i_{jk}, furthermore Riemann curvature tensor $R^p_{ijk} = (R_{pijk})$.

Thus, the geometry of space is determined by metric tensor g_{ij}. Riemann curvature tensor is represented as follows:

$$R_{\mu\nu kl} = \frac{1}{2} \cdot \left(g_{\mu l,\nu k} - g_{\nu l,\mu k} - g_{\mu k,\nu l} + g_{\nu k,\mu l} \right) + \Gamma_{\beta\mu l}\Gamma^\beta_{\mu k} - \Gamma_{\beta\mu k}\Gamma^\beta_{\nu l}$$

(3.4)

$$\Gamma^r_{ij} = \frac{1}{2} \cdot g^{rk} \left(g_{jk,i} + g_{ki,j} - g_{ij,k} \right)$$

(3.5)

Here, we use the notation $g_{jk,i}$ for $\dfrac{\partial g_{jk}}{\partial x^i}$.

As described above, Riemann curvature tensor $R_{\mu\nu kl}$ consists of fundamental metric tensor $g_{\mu\nu}$, therefore the structure of space-time is determined by metric tensor $g_{\mu\nu}$.

The solution of metric tensor $g_{\mu\nu}$ is found by gravitational field equation described above as the following:

$$R^{ij} - \frac{1}{2} \cdot g^{ij} R = -\frac{8\pi G}{c^4} \cdot T^{ij}$$

(3.6)

Furthermore, we have the following relation for scalar curvature R:

$$R = R^\alpha{}_\alpha = g^{\alpha\beta} R_{\alpha\beta}, \quad R^{\mu\nu} = g^{\mu\alpha} g^{\nu\beta} R_{\alpha\beta}, \quad R_{\alpha\beta} = R^j{}_{\alpha j\beta} = g^{ij} R_{i\alpha j\beta}$$

(3.7)

Ricci tensor is represented by

$$R_{\mu\nu} = \Gamma^\alpha_{\mu\alpha,\nu} - \Gamma^\alpha_{\mu\nu,\alpha} - \Gamma^\alpha_{\mu\nu}\Gamma^\beta_{\alpha\beta} + \Gamma^\alpha_{\mu\beta}\Gamma^\beta_{\nu\alpha} \quad (= R_{\nu\mu})$$

(3.8)

If the curvature of space is very small, the term of higher order than the second can be neglected and Ricci tensor becomes

$$R_{\nu\nu} = \Gamma^\alpha_{\mu\alpha,\nu} - \Gamma^\alpha_{\mu\nu,\alpha} \cdot$$

(3.9)

From Eq.(3.7), the major curvature of Ricci tensor ($\mu = \nu = 0$) is calculated as follows:

$$R^{00} = g^{00}g^{00}R_{00} = -1 \times -1 \times R_{00} = R_{00} \tag{3.10}$$

As is well known, Riemannian geometry is a geometry deals with a curved Riemann space, therefore Riemann curvature tensor is the principal quantity. All components of Riemann curvature tensor are zero for flat space and non-zero for curved space. If an only non-zero component of Riemann curvature tensor exists, the space is not flat space but curved space. Further, in a curved space, it is well known that the result of the parallel displacement of vector depends on the choice of the path and also the components of vector differ from the initial value after the vector parallel displacement is performed along a closed curve until it returns to the starting point.

For the reference, the following equations in Table.1 are effective for the calculation of General Relativity.

Table.1: Equations effective for the calculation of General relativity

$$\Gamma_{mmn} = (\Gamma_{nmm}) = -\frac{1}{2}g_{mm,n}$$

$$\Gamma_{mmm} = \frac{1}{2}g_{mm,m}$$

$$\Gamma_{mnn} = (\Gamma_{nmn}) = \frac{1}{2}g_{nn,m}$$

$$\Gamma_{mnm} = (\Gamma_{mmn}) = \frac{1}{2}g_{mm,n}$$

$$\Gamma^m_{mn} = \Gamma^m_{nm} = \frac{g_{mm,n}}{2g_{mm}}$$

$$\Gamma^m_{mm} = \frac{g_{mm,m}}{2g_{mm}}$$

$$\Gamma^m_{nn} = -\frac{g_{nn,m}}{2g_{mm}}$$

$$other \ \Gamma_{\mu\nu\lambda} = 0 \ (\because g_{\mu\nu} = 0 \ \ \mu \neq \nu) \ \ other \ \Gamma^\mu_{\nu\lambda} = 0$$

$$R_{\mu\nu kl} = \frac{1}{2}(g_{\mu l,\nu k} - g_{\nu l,\mu k} - g_{\mu k,\nu l} + g_{\nu k,\mu l}) + \Gamma_{\beta\mu l}\Gamma^\beta_{\nu k} - \Gamma_{\beta\mu k}\Gamma^\beta_{\nu l}$$

OUTLINE OF QUANTUM FIELD THEORY

Quantum field theory originated in the 1920s from the problem of creating a quantum mechanical theory of the electromagnetic field. In 1926, Max Born, Pascual Jordan, and Werner Heisenberg constructed such a theory by expressing the field's internal degrees of freedom as an infinite set of harmonic oscillators and by employing the usual procedure for quantizing those oscillators. Max Plank observed the behavior at the atomic level of radiation and heat on matter. And he observed that the energy absorbed or emitted was contained in small, discrete energy packets called quanta. This theory assumed that no electric charges or currents were present, and today would be called a free field theory. It is a theoretical framework that unifies non-relativistic quantum mechanics with special relativity. The first reasonably complete theory of quantum electrodynamics, which included both the electromagnetic field and electrically charged matter (specifically, electrons) as quantum mechanical objects, was created by Paul Dirac [6]. This quantum field theory could be used to model important processes such as the emission of a photon by an electron dropping into a quantum state of lower energy, a process in which the number of particles changes - one atom in the initial state becomes an atom plus a photon in the final state [8].

Quantum electrodynamics (QED) is a relativistic quantum field theory of electrodynamics, which describes electromagnetic interactions, which menans that it basically describes how light and matter interact with each other. More specifically, it mathematically describes all phenomena involving electrically charged particles interacting by means of exchange of photons. In quantum electrodynamics, electromagnetic forces are the result of the exchange of virtual photons. It says that photons are virtual because they are not observed directly, rather they are exchanged between two charged particles. The momentum carried by photons causes a recoil between two electrons giving rise to a repulsive force. It is extremely accurate predictions of quantities like the anomalous magnetic moment of the electron, and the Lamb shift of the energy levels of hydrogen. In technical terms, QED can be described as a perturbation theory of the electromagnetic quantum vacuum [7].

From the relativistic theory, the electromagnetic field tensor is given by

$$F^{\mu\nu} = \partial^{\mu} A^{\nu} - \partial^{\nu} A^{\mu} = \begin{pmatrix} 0 & -E_x & -E_y & -E_z \\ E_x & 0 & -B_z & B_y \\ E_y & B_z & 0 & -B_x \\ E_z & -B_y & B_x & 0 \end{pmatrix},$$

(4.1)

where A^{μ} is a 4-vector whose time component is the scalar potential and whose special component is the vector potential of electromagnetic field.

From which, the inhomogeneous Maxwell's equations can be written as

$$\partial_{\mu} F^{\mu\nu} - J^{\nu} = 0,$$

(4.2)

where J^{ν} are current densities.

For an electron coupled to the electromagnetic field, the Lagrangian density, from which everything can be derived, yields

$$L = -\frac{1}{4} F_{\mu\nu} F^{\mu\nu} - \bar{\psi}_e \gamma^{\mu} (\partial_{\mu} + ieA) \psi_e - m_e \bar{\psi}_e \psi_e,$$

(4.3)

where m_e is the rest mass of the electron and ψ_e is its field. It is expressed whthin the relastivic formalism that uses space-time four-coordinates numbered as $\mu, \nu = 0, 1, 2, 3$.

The Lagrangian density describes the free electromagnetic field defined by the four-ptential $A^{\mu} \equiv (\varphi, A)$, containing the Coulomb potential φ and vector potential A. The term γ^{μ} is the Dirac 4×4 matrices defined by

$$\gamma^0 = -i \begin{pmatrix} 0 & 1 \\ 1 & 0 \end{pmatrix}, \; \gamma = -i \begin{pmatrix} 0 & \sigma \\ \sigma & 0 \end{pmatrix}, \; \gamma_5 = \begin{pmatrix} 1 & 0 \\ 0 & -1 \end{pmatrix}.$$

(4.4)

QED, a quantum theory of electrons, positrons, and the electromagnetic field, was the first satisfactory quantum description of a physical field and of the creation and annihilation of quantum particles. QED involves a covariant and gauge invariant prescription for the calculation of observable quantities. Feynman's mathematical technique, based on his diagrams, initially seemed very different from the field-theoretic, operator-based approach of Schwinger and Tomonaga, but Freeman Dyson later showed that the two approaches were equivalent. The renormalization procedure for eliminating the awkward infinite predictions of quantum field theory was first implemented in QED.

QED has served as the model and template for all subsequent quantum field theories. Physically, QED describes charged particles (and their antiparticles) interacting with each other by the exchange of photons and the magnitude of these interactions can be computed using perturbation theory; these rather complex formulas have a remarkable pictorial representation as Feynman diagrams, which represent systematic pertubative expansion in power of fine-structure constant $\alpha = e^2 / \hbar c = 1/137$. QED was the theory to which Feynman diagrams were first applied. These diagrams were invented on the basis of Lagrangian mechanics. Using a Feynman diagram, one decides every possible path between the start and end points. Each path is assigned a complex-valued probability amplitude, and the actual amplitude we observe is the sum of all amplitudes over all possible paths, where the interaction of particles can be described.

The basic ingedient is the emission and absorption od a photon by an electron (or a positoron). This is all there is to describe all electromagnetic interaction in the QED.

According to QED, it can be seen that the quantum vacuum is filled with the zero-point electromagnetic field as shown in Fig. **3**. However this electromagnetic field is in the state of non-radiating mode and we cannot recognize the influence of zero point fluctuation of quantum electromagnetic field.

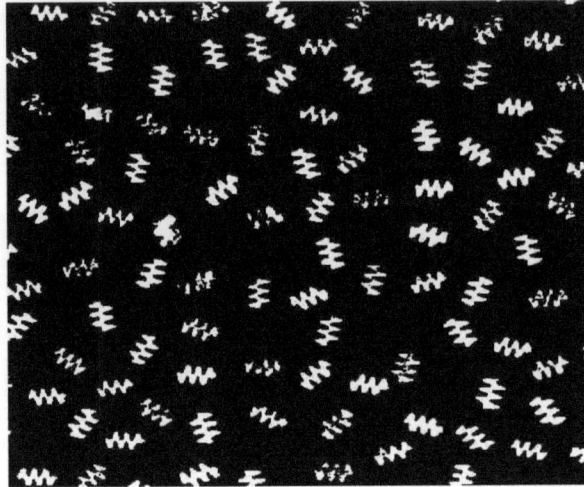

Figure 3: Zero-point fluctuation of Electromagnetic energy in a vacuum

In physics, the zero-point energy is the lowest possible energy that a quantum mechanical physical system may have and is the energy of the ground state.

All quantum mechanical systems have a zero-point energy. The term arises commonly in reference to the ground state of the quantum harmonic oscillator and its null oscillations. Zero-point energy is sometimes used as a synonym for the vacuum energy, an amount of energy associated with the vacuum of empty space. When the term is used in this way, sometimes it is referred to as the quantum vacuum zero-point energy. In cosmology, the vacuum energy is one possible explanation for the cosmological constant. The variation in zero-point energy as the boundaries of a region of vacuum move leads to the Casimir effect, which is observable in nano-scale devices.

Max Planck derived the formula for the average energy of a single energy radiator as

$$\varepsilon = \frac{h\nu}{e^{h\nu/kT} - 1},$$ (4.5)

where h is Planck's constant, ν is the frequency, k is Boltzmann's constant, and T is the absolute temperature.

Using this formula as a basis, Albert Einstein and Otto Stern published a paper of great significance in which they suggested for the first time the existence of a residual energy that all oscillators have at absolute zero. They called this "residual energy", which later became translated as zero-point energy. They carried out an analysis of the specific heat of hydrogen gas at low temperature, and concluded that the data are best represented if the vibrational energy is taken to have the form as

$$\varepsilon = \frac{h\nu}{e^{h\nu/kT} - 1} + \frac{h\nu}{2}.$$ (4.6)

According to this expression, an atomic system at absolute zero retains an energy of

$$E = \frac{\hbar\omega}{2}.$$ (4.7)

The energy of a system is relative, and is defined only in relation to some given state (which is often called the reference state). One might associate a motionless system with zero energy, but doing so is purely arbitrary.

In quantum physics, it is natural to associate the energy with the expectation value of a certain operator, the Hamiltonian of the system. For almost all quantum-mechanical systems, the lowest possible expectation value of this operator, which would be the zero-point energy, is not zero. Adding an arbitrary constant to the Hamiltonian gives an equivalent description of the physical system, but can make the zero-point energy different. Regardless of what constant is added to the Hamiltonian, the minimum momentum is always the same non-zero value.

For a one-dimensional harmonic oscillator of unit mass, the quantum mechanical Hamiltonian may be written as

$$H = \frac{1}{2}(\hat{p}^2 + \omega^2 \hat{q}^2),$$ (4.8)

where $\hat{p}$ and $\hat{q}$ are momentum and position operators.

From which, the Hamilton operator becomes

$$H = \hbar\omega\left(\hat{N} + \frac{1}{2}\right) = \hbar\omega(\hat{a}^\dagger \hat{a} + 1/2),$$ (4.9)

where $\hat{a}$ and $\hat{a}^\dagger$ are annihilation and creation operators respectively.

Thus the ground state energy of the quantum oscillator has the energy $\hbar\omega/2$ and the zero-point energy is the expectation value of the Hamiltonian of the system.

In quantum field theory, the fabric of space is visualized as consisting of fields, with the field at every point in space and time being a quantized simple harmonic oscillator, with neighboring oscillators interacting and one has a contribution of from every point in space, resulting in a calculation of infinite zero-point energy.

Here the zero-point energy is again the expectation value of the Hamiltonian; which is called the vacuum energy.

Casimir Effect

A phenomenon that is commonly presented as evidence for the existence of zero-point energy in vacuum is the Casimir effect. This effect was proposed in 1948 by Dutch physicist, Hendrik B. G. Casimir, who considered the quantized electromagnetic field between a pair of grounded, neutral metal plates.

The vacuum energy contains contributions from all wavelengths, except those excluded by the spacing between plates. As the plates draw together as shown in Fig. **4**, more wavelengths are excluded and the vacuum energy decreases.

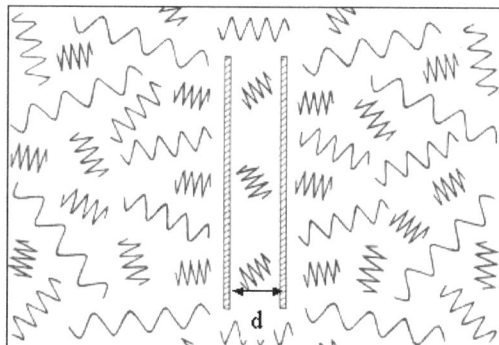

Figure 4: Zero-point energy between parallel plates

We consider thin plates at $x = 0$, $x = L$ and $y = 0$, $y = L$ to form a box.

A conducting plate imposes boundary conditions on the radiation field and the sum of zero-point energies given by[14]

$$U = \sum{}' \frac{1}{2} \hbar \omega \times 2 = \sum_{n_x, n_y, n_z}{}' \hbar c \left[\left(\frac{\pi n_x}{L} \right)^2 + \left(\frac{\pi n_y}{L} \right)^2 + \left(\frac{\pi n_z}{L} \right)^2 \right]^{1/2} . \qquad (4.10)$$

The summation Σ' means that whenever one of n's vanishes, we drop the multiplicity 2 for possible transverse polarizations.

We regard the size of the plate to be large $L \to \infty$, we can replace the summation by the term of integrals over $k_{x,y} = \pi n_{x,y} / L$ given by

$$U = \left(\frac{L}{\pi} \right)^2 \sum_{n_z}{}' \int_0^\infty dk_x \int_0^\infty dk_y \, \hbar c \left[k_x^2 + k_y^2 + \left(\frac{\pi n_z}{d} \right)^2 \right] . \qquad (4.11)$$

If we consider the case when the plates don't exist, the energy yield the zero-point energy density multiplied by the volume $L^2 d$ shown as

$$U_0 = L^2 d \int \frac{d\vec{k}}{(2\pi)^3} \hbar c \left[k_x^2 + k_y^2 + k_z^2 \right]^{1/2} = L^2 d \int_0^\infty \frac{dk_x}{\pi} \int_0^\infty \frac{dk_y}{\pi} \int_0^\infty \frac{dk_z}{\pi} \hbar c \left[k_x^2 + k_y^2 + k_z^2 \right]^{1/2} . \qquad (4.12)$$

What is observable is the difference of energy given by $U - U_0$, which can be obtained by utilizing the Euler-MacLaurin formula as

$$U - U_0 = -\frac{\pi^2 L^2 \hbar c}{720 d^3} , \qquad (4.13)$$

From which, there is a attractive force generated between two conductive plates per unit area as shown in Fig. **5** given by

$$F/A = -\frac{(U - U_0)'}{A} = -\frac{\pi^2 \hbar c}{240 d^4} , \qquad (4.14)$$

where A is an area of the metal plates.

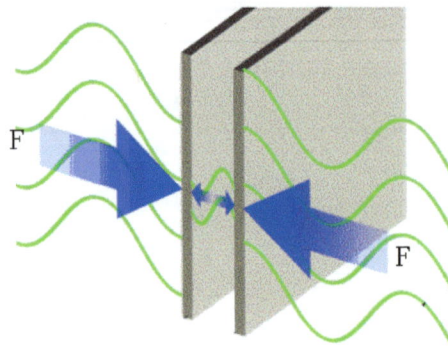

Figure 5: Casimir effect between plates

This force has been measured and found to be in good agreement with the theory [18]. In quantum field theory, the vacuum state (also called the vacuum) is the quantum state with the lowest possible energy. Generally, it contains no physical particles. The term "zero-point field" is sometimes used as a synonym for the vacuum state of an individual quantized field. According to present understanding of what is called the vacuum state or the quantum vacuum, there is no simple empty space, but the vacuum state is not truly empty but instead contains fleeting electromagnetic waves and particles that pop into and out of existence according to quantum mechanics. Thus the vacuum state is associated with a zero-point energy, and this zero-point energy has measurable effects. In physical cosmology, the energy of the vacuum state appears as the cosmological constant. In fact, the energy of a cubic centimeter of empty space has been calculated to be about one trillionth of an erg.

By utilizing the Casmir force, R.L.Forward proposed a vacuum-fluctuation battery as shown in Fig. **6** [2].

As shown in this figure, any pair of conducting plates as close distance experiences as attractive Casimir force due to the electromagnetic zero-point fluctuations of vacuum. By applying a charge of the same polarity to each conducting plate, a repulsive electrostatic force will be produced, which opposes the Casimir force. Then the plates will move toward each other and the Casimir force will add energy to the electric field between the plates. Thus the battery can be recharged by making the electrical forces slightly stronger than the Casimir force to re-expand the foliated conductor.

From the quantum field theory, the following important properties can be predicted which are different from classical theorems.

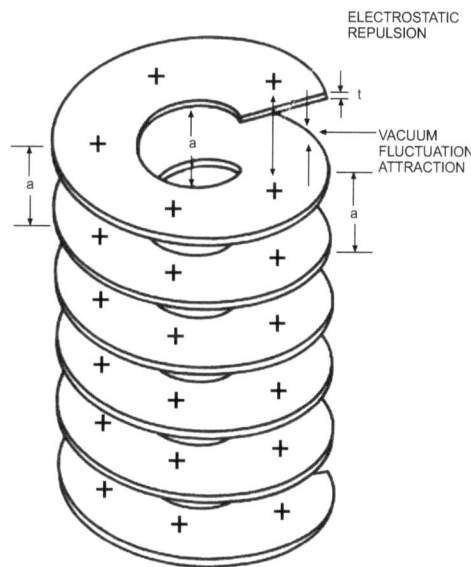

Figure 6: Vacuum-fluctuation battery proposed by Forward

Vacuum Polarization

The Dirac sea is a collection of infinite number of electrons in negative energy states. Even though it has an infinite negative charge, we cannot detect it because there is no preferred direction to produce an electric field.. If the homogeneity of it is broken, the distribution of the negative energy electrons is no longer homogeneous and can be physically observable. Supposing that the positive point charge is placed in the vacuum filled with negative energy electrons, they are attracted to the positive charge and screening the point charge and therefore the charge of the point charge would appear less than the original charge viewed from far away. In other words, the fine structure constant would appear larger when it is measured at smaller distant scales.

This effect is the vacuum polarization of causing the fine-structure constant to grow at higher momentum transfers. Related to this effect, Schwinger showed that the g-factor which shows the interaction of electrons with the magnetic field can be given by

$$g = 2\left(1 + \frac{\alpha}{2\pi}\right),$$

(4.15)

where α is a fine-structure constant.

Bose-Einstein Condensation

A Bose–Einstein condensate (BEC) is a state of matter of a dilute gas of weakly interacting bosons confined in an external potential and cooled to temperatures very near to absolute zero. Under such conditions, a large fraction of the bosons occupy the lowest quantum state of the external potential, at which point quantum effects become apparent on a macroscopic scale.

The slowing of atoms by use of cooling apparatus produces a singular quantum state known as a Bose condensate or Bose–Einstein condensate. This phenomenon was predicted in 1925 by generalizing Bose's work on the statistical mechanics of (massless) photons to (massive) atoms.

The result of the efforts of Bose and Einstein is the concept of a Bose gas, governed by Bose–Einstein statistics, which describes the statistical distribution of identical particles with integer spin, now known as bosons. Bosonic particles, which include the photon as well as atoms such as helium-4, are allowed to share quantum states with each other. Einstein demonstrated that cooling bosonic atoms to a very low temperature would cause them to fall (or "condense") into the lowest accessible quantum state, resulting in a new form of matter.

This transition occurs below a critical temperature, which for a uniform three-dimensional gas consisting of non-interacting particles with no apparent internal degrees of freedom is given by:

$$T_c = \left(\frac{n}{\zeta(3/2)}\right)^{2/3} \frac{2\pi\hbar}{mk_B},$$

(4.16)

where T_c is the critical temperature, m is the mass of the boson and n is the density of particles.

Consider now a gas of particles, which can be in different momentum states labeled. If the number of particles is less than the number of thermally accessible states, for high temperatures and low densities, the particles will all be in different states. In this limit, the gas is classical. As the density increases or the temperature decreases, the number of accessible states per particle becomes smaller, and at some point more particles will be forced into a single state than the maximum allowed for that state by statistical weighting. From this point on, any extra particle added will go into the ground state.

Usually Bose-Einstein condensation occurs when matter is cooled down, and there are just not enough many sates for all the particles to fit. For being bosons, which can squeeze into the same state, any excess bosons end up squeezing into the same, ground state. For photon gas, they cannot go under B-E condensation because they just get absorbed by the material they are un contact with. However, they do for a laser, where the photons all have the same wave-function.

The squeezed coherent state is any state of the quantum mechanical Hilbert space such that the uncertainty principle is saturated. That is, the product of the corresponding two operators takes on its minimum value. The simplest such state is the ground state of the quantum harmonic oscillator. The next simple class of states that satisfies this identity are the family of coherent states.

For the coherent state, the number-phase uncertainty with minimum uncertainty state can be given by

$$\Delta N \cdot \Delta\varphi = 1/2,$$

(4.17)

where N is a number of the particle and φ is their phase.

For the squeezed state, ΔN gets smaller and $\Delta \varphi$ becomes large still satisfying the relation $\Delta N \cdot \Delta \varphi = 1/2$.

The idea behind this is that the circle denoting a coherent state in a quadrature diagram (see Fig. **7**) has been "squeezed" to an ellipse of the same area.

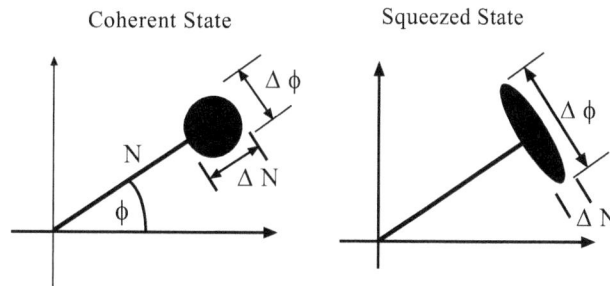

Figure 7: Two quantum states of bosons

Superstring Theory

The ultimate step for quantum field theory further from the quantum field theory is the unification theory known as super string theory [16]. It was originally proposed as a theory of the strong interaction, and the basic idea is that the fundamental objects in the universe are not point-like elementary particles, but are instead objects spread out in one-dimensional called strings[5]. Excitations of the string give the different particles in the universe. The basic idea of the super string is that the fundamental constitutes of reality are strings of the Plank length rather than point particles, which vibrate at resonant frequencies. The tension in a string is on the order of 10^{44} Newtons. The graviton which is the proposed messenger particle of the gravitational force is predicted by the theory to be a string with wave amplitude zero.

Superstring theory is based on one principle that everything at its most microscopic level consists of combinations of vibrating strands. Superstring theory provides a single explanatory framework capable of encompassing all forces and all matter.

The string is a one-dimensional object, a mathematical curve. There are two kinds of strings, that is, open string, which have endpoints, and closed string, which from a topological viewpoint are circles (see Fig. **8**). The open string has two special points, the endpoints. It is possible to assume that the open string carries 'charge' at its endpoints. For instance, there is a 'quark' at one end of the string and an 'antiquark' at the other end shown in Fig. **8** (**a**).

The closed string is a loop with no 'free ends', topologically equivalent to circles as shown in Fig. **8** (**b**). The fundamental size of string, called string length l_s, is Planck length (10^{-35}m).

The different vibrational patterns of a fundamental string give rise to different masses and force charges. According to superstring theory, the properties of an elementary particle (its mass and its various force charges) are determined by the precise resonant pattern of vibration that its internal string executes. The energy of a particular vibrational string pattern depends on its amplitude and its wavelength. The greater the amplitude and the shorter the wavelength, string has the greater the energy.

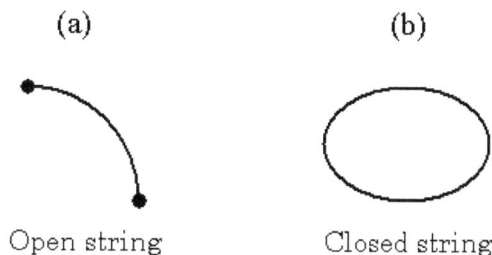

(a)   (b)

Open string   Closed string

Figure 8: Open and closed strings

Thus, according to superstring theory, the mass of an elementary particle is determined by the energy of the vibrational pattern of its internal string. Heavier particles have internal strings that vibrate more energetically, while lighter particles have internal strings that vibrate less energetically.

Each elementary particle is composed of a single string, and all strings are absolutely identical. Differences between the particles arise because their respective strings undergo different resonant vibrational patterns. As strings twist and vibrate while meandering through the extended and curled-up dimensions, a small subset of their vast oscillatory repertoire consists of vibrations with spin equal to 1 or 2. These are the candidate force-carrying string-vibrational states. There is always one vibrational pattern that is massless and has spin-2, *i.e.* graviton.

In General Relativity, the gravity is described by a massless field of spin-2, the graviton field. The occurrence of a massless spin-2 particle in the spectrum of states of the closed string suggests an alternative use for superstring theory as a possible framework for gravitation.

By the way, the Yang-Mills field is a massless field of spin-1, the gauge field. The occurrence of a massless spin-1 particle (gauge particles) in the spectrum of states of the open string suggests an alternative use for superstring theory as a possible framework for gauge theory.

As described above, to constitute the fundamental element in nature by the extended nature of the string not a point particle eliminates the problem of ultraviolet divergence. The extended nature of the string is crucial to its ability to resolve the conflicts between General Relativity and Quantum Mechanics that a point particle theory cannot. Superstring theory is the most promising candidate for a quantum theory of all known interactions (*i.e.*, the electromagnetic interaction, the strong interaction, the weak interaction and the gravitational interaction).

The mass of the string at the fundamental energy state can be given by

$$m = m_0 \sqrt{1 + (D-2)\left(\left(\sum_{n=1}^{\infty} n\right) - \sum_{n=1}^{\infty}\left(n - \frac{1}{2}\right)\right)} ,$$ (4.18)

where D is a dimension of the Universe.

As $\sum_{n=1}^{\infty} n = \zeta(-1) = 1/12$ and $\sum_{n=1}^{\infty}(n - 1/2) = 1/24$, it is estimated that the dimensions space-time manifold, but it includes more than 4 dimensions. Additional dimensions required by the superstring theory are confined in a very small manifold with the size on the order of the Plank length.

Superstring theory is not the first theory to propose the extra dimensions. The Kaluza-Klein theory proposed the extra dimensions to combine electromagnetic field and the gravitational field.

By this theory, singularities are avoided because interactions are spread out and the divergence associated with gravitational interaction disappear, instead of the point particles that cause the theory blow up, that is to get calculations with infinite result.

By this theory, the quantum of the gravitational field will be a massless spin-2 particle.

CHAPTER 4

Field Propulsion Systems Based on the Advanced Physics

Abstract: This chapter consists of two sections, the first section is for the field propulsion systems based on the General Relativity Theory, such as space drive propulsion and warp drive propulsion. The second section is for the systems on the Quantum Field Theory, which are based on the hypothesis that the inertial mass can be developed from the interaction of elementary particles with the vacuum electromagnetic zero-point field. Several field propulsion systems on these theories are presented in this chapter.

INTRODUCTION

Field propulsion is the concept of propulsion theory of spaceship not based on usual momentum thrust but based on momentum derived from an interaction of the spaceship with external fields. Concepts for field propulsion that rely on physics outside the present paradigms hold various schemes for advanced space propulsion. In this chapter, the basic concepts for such means of propulsion within the present physics paradigm are presented.

GENERAL RELATIVISTIC FIELD PROPULSION

Based on General Relativity, this section describes the behavior of space. This notion regarding the space is basic foundation for investigating field propulsion.

Fundamental Supposition of Space

On the supposition that space is an infinite continuum, continuum mechanics can be applied to the so-called "vacuum" of space. This means that space can be considered as a kind of transparent elastic field. That is, space as a vacuum performs the motion of deformation such as expansion, contraction, elongation, torsion and bending. The latest expanding universe theory (Friedmann, de Sitter, inflationary cosmological model) supports this assumption. We can regard space as an infinite elastic body like rubber.

If space curves, then an inward normal stress "$-P$" is generated. This normal stress, *i.e.* surface force serves as a sort of pressure field.

$$-P = N \cdot (2R^{00})^{1/2} = N \cdot (1/R_1 + 1/R_2) \,, \tag{1.1}$$

where N is the line stress, R_1, R_2 are the radius of principal curvature of curved surface, and R^{00} is the spatial curvature.

A large number of curved thin layers form the unidirectional surface force, *i.e.* acceleration field. Accordingly, the spatial curvature R^{00} produces the acceleration field α.

The fundamental three-dimensional space structure is determined by quadratic surface structure. Therefore, Gaussian curvature K in two-dimensional Riemann space becomes important. The relationship between K and the major component of spatial curvature R^{00} is given by

$$K = \frac{R_{1212}}{(g_{11}g_{22} - g_{12}{}^2)} = \frac{1}{2} \cdot R^{00} \,, \tag{1.2}$$

where R_{1212} is non-zero component of Riemann curvature tensor.

Applying membrane theory, the following equilibrium conditions are obtained in quadratic surface, given by

$$N^{\alpha\beta} b_{\alpha\beta} + P = 0 \,, \tag{1.3}$$

Curvature of space plays a significant role for propulsion theory.
If space curves, then inward normal stress (surface force) "P" is
generated ⇒ A Sort of Pressure Field

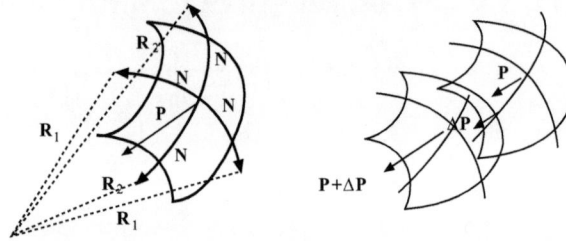

A large number of curved thin layers form
the unidirectional surface force, i.e.,
Acceleration Field α.

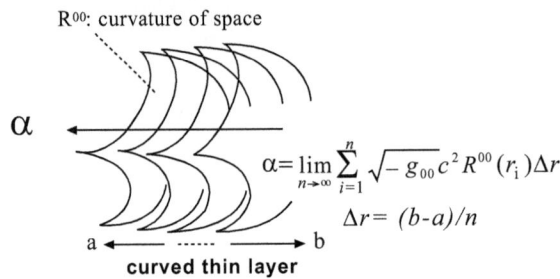

Figure 1: Curvature of Space

where $N^{\alpha\beta}$ is a membrane force, *i.e.* line stress of curved space, $b_{\alpha\beta}$ is second fundamental metric of curved surface, and P is the normal stress on curved surface [8].

The second fundamental metric of curved space $b_{\alpha\beta}$ and principal curvature $K_{(i)}$ have the following relationship using the metric tensor $g_{\alpha\beta}$,

$$b_{\alpha\beta} = K_{(i)}g_{\alpha\beta} \,. \tag{1.4}$$

Therefore we get

$$N^{\alpha\beta}b_{\alpha\beta} = N^{\alpha\beta}K_{(i)}g_{\alpha\beta} = g_{\alpha\beta}N^{\alpha\beta}K_{(i)} = N_\alpha{}^\alpha K_{(i)} = N \cdot K_{(i)} \tag{1.5}$$

From Eq.(1.3) and Eq.(1.5), we get

$$N_\alpha{}^\alpha K_{(i)} = -P \,. \tag{1.6}$$

As for the quadratic surface, the indices α and i take two different values, *i.e.* 1 and 2, therefore Eq.(1.6) becomes

$$N_1^1 K_{(1)} + N_2^2 K_{(2)} = -P \,, \tag{1.7}$$

where $K_{(1)}$ and $K_{(2)}$ are principal curvature of curved surface and are inverse number of radius of principal curvature (*i.e. $1/R_1$* and *$1/R_2$*).

The Gaussian curvature K is represented as

$$K = K_{(1)} \cdot K_{(2)} = (1/R_1) \cdot (1/R_2) \,. \tag{1.8}$$

Accordingly, suppose $N_1^{\ 1} = N_2^{\ 2} = N$, we get

$$N \cdot (1/R_1 + 1/R_2) = -P . \tag{1.9}$$

It is now understood that the membrane force on the curved surface and each principal curvature generate the normal stress "$-P$" with its direction normal to the curved surface as a surface force. The normal stress $-P$ is towards the inside of surface as showing in Fig. **1**. A thin-layer of curved surface will be taken into consideration within a spherical space having a radius of R and the principal radii of curvature which are equal to the radius ($R_1 = R_2 = R$). From Eqs. (1.2) and (1.8), we get

$$K = \frac{1}{R_1} \cdot \frac{1}{R_2} = \frac{1}{R^2} = \frac{R^{00}}{2} . \tag{1.10}$$

Considering $N \cdot (2/R) = -P$ of Eq.(1.9), and substituting Eq.(1.10) into Eq.(1.9), the following equation is obtained:

$$-P = N \cdot \sqrt{2R^{00}} . \tag{1.11}$$

Since the membrane force N (serving as the line stress) can be assumed to have a constant value, Eq.(1.11) indicates that the curvature R^{00} generates the inward normal stress P of the curved surface. The inwardly directed normal stress serves as a kind of pressure field. When the curved surfaces are included in a great number, some type of unidirectional pressure field is formed. A region of curved space is made of a large number of curved surfaces and they form the field of unidirectional surface force (*i.e.* normal stress). Since the field of surface force is the field of a kind of force, matter in the field is accelerated by the force, *i.e.* we can regard the field of surface force as the acceleration field. Accordingly, the curvature R^{00} produces the acceleration field α .

Therefore, the curvature of space plays a significant role.

Other Standpoint of View of Space (Fundamental Structure Equation of Space)

The mechanical structure due to space strain is described here from another point of view.

As shown in Fig. **2**, if the line element between the arbitrary two near points (A and B) in space region **S** (before structural deformation) is defined as $ds = g_i dx^i$, the infinitesimal distance between the two near points is given by

$$ds^2 = g_{ij} dx^i dx^j . \tag{1.12}$$

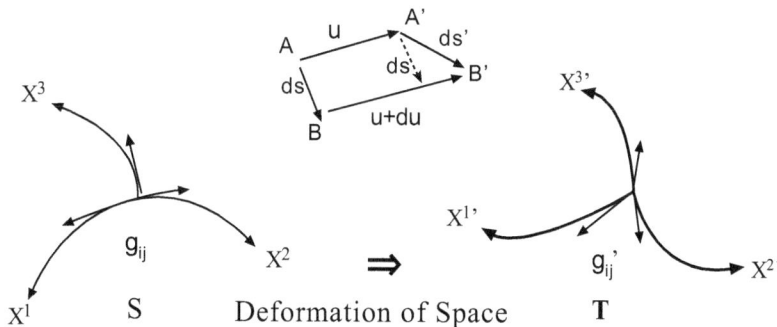

Figure 2: Fundamental structure of Space

Let us assume that a space region **S** is structurally deformed by external physical action and transformed to space region **T**. In the deformed space region **T**, the line element between the identical two near point (A'and B') of the identical space region newly changes, differs from the length and direction, and becomes $ds' = g_i' dx^i$. Therefore, the infinitesimal distance between the two near points using the convected coordinate ($x'^i = x^i$) is given by

$$ds'^2 = g'_{ij}dx^i dx^j \, . \tag{1.13}$$

The g'_i is the transformed base vector from the original vase vector g^i and the g'_{ij} is the transformed metric tensor from the original metric tensor g_{ij}. Since the degree of deformation can be expressed as the change of distance between the two points, we get

$$ds'^2 - ds^2 = g'_{ij}dx^i dx^j - g_{ij}dx^i dx^j = (g'_{ij} - g_{ij})dx^i dx^j = r_{ij}dx^i dx^j \tag{1.14}$$

Hence the degree of geometrical and structural deformation can be expressed by the quantity denoted change of metric tensor, *i.e.*,

$$r_{ij} = g'_{ij} - g_{ij} \, . \tag{1.15}$$

On the other hand, the state of deformation can be also expressed by the displacement vector "*u*".

From the continuum mechanics [8,14], using the following equations;

$$du = g^i u_{i:j}dx^j \, , \tag{1.16}$$

$$ds' = ds + du = ds + g^i u_{i:j}dx^j \, . \tag{1.17}$$

We use the usual notation ":" for covariant differentiation. From the usual continuum mechanics, the infinitesimal distance after deformation becomes

$$ds'^2 - ds^2 = r_{ij}dx^i dx^j = (u_{i:j} + u_{j:i} + u^k_{\ :i}u_{k:j})dx^i dx^j \tag{1.18}$$

The terms of higher order than second $u^k_{\ :i}u_{k:j}$ can be neglected if the displacement is enough small value. As the actual physical space can be dealt with the minute displacement from the trial calculation of strain, we get

$$r_{ij} = u_{i:j} + u_{j:i} \, . \tag{1.19}$$

Whereas, according to the continuum mechanics, the strain tensor e_{ij} is given by

$$e_{ij} = \frac{1}{2} \cdot r_{ij} = \frac{1}{2} \cdot (u_{i:j} + u_{j:i}) \, . \tag{1.20}$$

So, we get

$$ds'^2 - ds^2 = (g'_{ij} - g_{ij})dx^i dx^j = 2e_{ij}dx^i dx^j \, , \tag{1.21}$$

where g'_{ij}, g_{ij} is a metric tensor, e_{ij} is a strain tensor, and $ds'^2 - ds^2$ is the square of the infinitesimal distance between two infinitely proximate points x^i and $x^i + dx^i$.

Eq.(1.21) indicates that a certain geometrical structural deformation of space is shown by the concept of strain. In a word, the change of metric tensor $(g'_{ij} - g_{ij})$ due to the existence of mass energy or electromagnetic energy tensor produces the strain field e_{ij}.

This means that a certain geometrical structural deformation of space is described by the concept of strain. Since the space-time is distorted, the infinitesimal distance between two infinitely proximate points x^i and $x^i + dx^i$ is important for geometry of the space-time; the physical strain is generated by the difference of geometrical metric of space-

space-time. Namely, a certain structural deformation is described by strain tensor e_{ij}. From Eq.(1.21), the strain of space is described as follows:

$$e_{ij} = 1/2 \cdot (g_{ij}' - g_{ij}) \tag{1.22}$$

As is well known in the continuum mechanics, the elastic force F^i is given by the gradient of stress tensor σ^{ij}, and using elastic law with the elastic modulus $E^{ij\mu\nu}$, if we apply the continuum mechanics to above result, the strain tensor e_{ij} produces the stress field σ^{ij} and from the equilibrium conditions of continuum, we get

$$-F^i = \sigma^{ij}{}_{:j} \quad \text{and} \quad \sigma^{ij} = E^{ij\mu\nu} e_{\mu\nu}. \tag{1.23}$$

The stress tensor σ^{ij} is a surface force and F^i is a body force. The body force is an equivalent gravitational action because of acting all elements of space uniformly. From Eq.(1.23), the following equation is obtained

$$-F^i = \sigma^{ij}{}_{:j} = (E^{ij\mu\nu} e_{\mu\nu})_{:j} = E^{ij\mu\nu} e_{\mu\nu:j}. \tag{1.24}$$

Here, we assumed that $E^{ij\mu\nu}$ is constant for covariant differentiation.

Further more, expanding the concept of vector parallel displacement in Riemann space, the Eq.(1.24) is also changed to the following equation (see APPENDIX.A)

$$F^i = E^{ij\mu\nu} R_{\mu\nu kl:j} dA^{kl}. \tag{1.25}$$

Eq.(1.25) indicates the same concept of curved space and yield the elastic force F which produces the acceleration α of space drive. We can say again that the curvature of space plays a significant role.

The non-zero component of Eq.(1.25) is just only one equation as follows:

$$F^3 = F = E^{3330} (R_{3030} A^{30})_{:3} = E^{3330} \cdot \partial (R_{3030} A^{30}) / \partial r. \tag{1.26}$$

As described above, an important analytical method relating the concept of continuum mechanics as a deformation with the concept of Riemannian geometry is the concept of the parallel displacement of vector.

All kinds of new researches rely on the fine-structure of space as a vacuum. It is also worth noting that this result yields the principle of constancy of light velocity in Special Relativity (see APPENDIX.B).

General Relativity Solutions for Field Propulsion

From the following linear approximation scheme for the gravitational field equation 1) weak gravitational field, *i.e.* small curvature limit, 2) quasi-static, 3) slow-motion approximation (*i.e.* $v/c < 1$), we get the following relation between acceleration of curved space and curvature of space:

$$\alpha^i = \sqrt{-g_{00}} c^2 \int_a^b R^{00}(x^i) dx^i, \tag{1.27}$$

where α^i: acceleration(m/s^2), g_{00}: time component of metric tensor, a-b: range of curved space region(m), x^i: components of coordinate (i=0,1,2,3), c: velocity of light, R^{00}: major component of spatial curvature (see APPENDIX.C).

Eq.(1.27) indicates that the acceleration field α^i is produced in curved space. As is well known in General Relativity, in the curved space region, the massive body "m(kg)" existing in the acceleration field is subjected to the following force $F^i(N)$,

$$F^i = m\Gamma^i_{jk} \cdot \frac{dx^j}{d\tau} \cdot \frac{dx^k}{d\tau} = m\sqrt{-g_{00}}\, c^2 \Gamma^i_{jk} u^j u^k = m\alpha^i ,$$

(1.28)

where u^j, u^k are the four velocity, Γ^i_{jk} is the Riemannian connection coefficient, and τ is the proper time.

Eq.(1.28) yields more simple equation from above-stated linear approximation,

$$F^i = m\sqrt{-g_{00}}\, c^2 \Gamma^i_{00} = m\alpha^i = m\sqrt{-g_{00}}\, c^2 \int_a^b R^{00}(x^i)dx^i .$$

(1.29)

Setting $i=3$ (*i.e.* direction of radius of curvature:r), we get the following equation as Newton's second law

$$F^3 = F = m\alpha = m\sqrt{-g_{00}}\, c^2 \int_a^b R^{00}(r)dr = m\sqrt{-g_{00}}\, c^2 \Gamma^3_{00} .$$

(1.30)

The acceleration (α) of curved space and its Riemannian connection coefficient (Γ^3_{00}) are given by

$$\alpha = \sqrt{-g_{00}}\, c^2 \Gamma^3_{00} , \qquad \Gamma^3_{00} = \frac{-g_{00,3}}{2g_{33}} ,$$

(1.31)

where c: velocity of light, g_{00} and g_{33}: component of metric tensor, $g_{00,3}$: $\partial g_{00}/\partial x^3 = \partial g_{00}/\partial r$.

We choose the spherical coordinates " $ct=x^0$, $r=x^3$, $\theta=x^1$, $\varphi=x^2$ " in space-time. The acceleration α is represented by the equation both in the differential and in the integral form. Practically, since the metric is usually given, the differential form has been found to be advantageous.

Next, we expand on these categories as they relate to other solutions of gravitational field equation, that is, the concrete acceleration α is derived from Eq.(1.31).

External Schwarzschild Solution

The metrics are given by:

$$g_{00} = -(1-r_g/r), g_{11} = g_{22} = 1, g_{33} = 1/(1-r_g/r),$$
$$and \ \ other \ g_{ij} = 0 .$$

(1.32)

where r_g is the gravitational radius (*i.e.* $r_g = 2GM/c^2$). Combining Eq.(1.31) with Eq.(1.32) yields:

$$\alpha = G \cdot \frac{M}{r^2}, (r_g \langle r) \quad ,$$

(1.33)

where G is a gravitational constant and M is a total mass.

Reissner-Nordstrom Charged Mass Solution

The metrics outside of charged and spherically symmetric mass are given by:

$$g_{00} = -(1-r_g/r+Q^2/r^2), g_{11} = g_{22} = 1, g_{33} = 1/(1-r_g/r+Q^2/r^2),$$
$$and \ \ other \ g_{ij} = 0.$$

(1.34)

where $Q^2 = Gq^2/c^4$ ($q = electric\ charge$), $r_g = 2GM/c^2$.

Eq.(1.34) reduces to the Schwarzschild solution if electric charge "q" is zero.

Combining Eq.(1.31) with Eq.(1.34) yields:

$$\alpha = G \cdot \frac{M}{r^2} - \frac{Gq^2}{c^2 r^3} < G \cdot \frac{M}{r^2}, \ (r_g < r, Q^2 < r^2) \ . \tag{1.35}$$

Eq.(1.35) indicates that the electric charge weakens the gravitational acceleration.

Kerr Rotating Mass Solution

The metrics outside of spinning mass are given by:

$$g_{00} = -\left(1 - \frac{r_g r}{r^2 + h^2 \cos^2 \theta}\right) \ , \ g_{33} = \frac{r^2 + h^2 \cos^2 \theta}{r^2 - r_g r + h^2}, \tag{1.36}$$

where $h = J / Mc$ (J = *angular momentum*), $r_g = 2GM / c^2$.

Eq.(1.36) reduces to the Schwarzschild solution if the angular momentum "*J*" is zero.

Combining Eq.(1.31) with Eq.(1.36) yields:

$$\alpha = G \cdot \frac{M}{r^2} \cdot \frac{(1 - h^2 \cos^2 \theta / r^2)}{(1 + h^2 \cos^2 \theta / r^2)^3} < G \cdot \frac{M}{r^2} \ , \ (r_g < r, h^2 < r^2) \ . \tag{1.37}$$

Eq.(1.37) indicates that the rotation weakens the gravitational acceleration.

Internal Schwarzschild Solution

The space-time metrics inside of a static, constant energy density, perfect fluid sphere are given by:

$$g_{00} = -\left[\frac{3}{2} \cdot (1 - K\rho a^2 / 3)^{\frac{1}{2}} - \frac{1}{2} \cdot (1 - K\rho r^2 / 3)^{\frac{1}{2}}\right]^2, \ g_{33} = \frac{1}{1 - K\rho r^2 / 3} \ ,$$
$$g_{11} = g_{22} = 1, \ and \ other \ g_{ij} = 0 \tag{1.38}$$

where $K = 8\pi G / c^4$, ρ is the energy density (J/m^3), "a" is the radius of energy density (*i.e.* fluid boundary at $r=a$). This solution corresponds to the so-called Poisson equation. While, External Schwarzschild Solution corresponds to the so-called Laplace equation.

Combining Eq.(1.31) with Eq.(1.38) yields:

$$\alpha = \frac{GM}{a^3} \cdot r \ \ (r_g < a) \ . \tag{1.39}$$

Eq.(1.39) reduces to the Eq.(1.33), if $r=a$, and the continuity at "$r=a$" links the internal solution to external solution.

De Sitter Solution

In the latest cosmology, the terms vacuum energy and cosmological term "Λg^{ij}" are used synonimously. Λ is a constant known as the cosmological constant. The cosmological term is identical to the stress-energy associated with the vacuum energy. The properties of vacuum energy, *i.e.* cosmological term are crucial to expansion of the Universe, that is, to inflationary cosmology. The vacuum energy in de Sitter solution yields the result that the expansion accelerates with time and the total energy with a comoving volume grows exponentially. These facts are due to the elastic nature of vacuum and support the following basic concept: space is an infinite continuum.

Now, concerning the de Sitter cosmological model with non-zero vacuum energy (*i.e.* cosmological constant), the de Sitter line element is written as

$$ds^2 = -(1 - \frac{1}{3}\Lambda r^2)c^2 dt^2 + \frac{1}{1 - \frac{1}{3}\Lambda r^2} dr^2 + r^2(d\theta^2 + \sin^2\theta d\varphi^2) \,. \tag{1.40}$$

The metrics are given by

$$g_{00} = -(1 - 1/3 \cdot \Lambda r^2), \quad g_{11} = g_{22} = 1, \quad g_{33} = 1/(1 - 1/3 \cdot \Lambda r^2),$$
$$and \ \ other \ g_{ij} = 0 \tag{1.41}$$

The acceleration α of de Sitter solution can be obtained by combining Eq.(1.31) with Eq.(1.41)

$$\alpha = \frac{1}{3}c^2\Lambda r \quad (1 > 1/3 \cdot \Lambda r^2) \,. \tag{1.42}$$

The acceleration induced by cosmological constant is proportional to the distance "*r*" from the generative source.

Space Drive Propulsion System

Although several kinds of ideas regarding the field propulsion used General Relativity are introduced thereinafter sections, they are not considered well from the stand point of propulsion technology like electric propulsion and laser propulsion described in chapter 2. In this section, space drive propulsion system considered comparatively well is introduced firstly in detail.

Background

The principle of space drive propulsion system is derived from General Relativity and the theory of continuum mechanics. We assume the so-called "vacuum" of space as an infinite elastic body like rubber. The curvature of space plays a significant role for the propulsion theory. From the gravitational field equation, the strong magnetic field as well as mass density generates the curvature of space and this curved space region produces the uni-directional acceleration field. The spaceship in the curved space can be propelled in a single direction. Since the force they produce acts uniformly on every atom inside the spaceship, accelerations of any magnitude can be produced with no strain on the crews, that is, there is no action of inertial force because the thrust is a body force (*i.e.*, it is equivalent to free-fall).

Yoshinari Minami proposed new propulsion theory used General Relativity in 1988. The paper entitled "Space Strain Propulsion System" is presented at 16th ISTS 1988 [28]. After then, Minami presented the second paper entitled "Possibility of Space Drive Propulsion" at 45th IAF 1994 [25]. The term of "space strain" is changed to "space drive" receiving the recommendation by Robert L. Forward [11]. Minami derived the equation of curvature of space induced by magnetic field in 1988. It was found that this equation was accordance with the equation that Levi-Civita considered (*i.e.*, the static magnetic field creates scalar curvature) by Claudio Maccone and Minami in 1995 (APPENDIX.D).

In the beginning, the acceleration generated by curvature of space induced by strong magnetic field based on external and internal Schwarzschild solution was studied. However, superior acceleration based on de Sitter solution is obtained and presented at 47th IAF in 1996 [26]. *The acceleration derived from de Sitter solution does not require a strong magnetic field.* At the present day, space drive propulsion system based on de Sitter solution needs not strong magnetic field but the technology to excite space.

The details are published in JBIS, Vol.50 [27] and presented at STAIF-98, in 1998 [29]. The space drive propulsion system is also a field propulsion system utilizing the action of the medium of strained or deformed field of space, and is based on the propulsion principle of the kind of pressure thrust. The curvature of space plays a significant role for propulsion theory. The acceleration performance of this system is found by the solution of the gravitational field equation, such as the Schwarzschild solution, Reissner-Nordstrom solution, Kerr solution, and de Sitter solution. The concept and the details of this propulsion system are described below.

Space Drive Propulsion Theory

The theory of space drive propulsion is summarized as follows.

1) On the supposition that space is an infinite continuum, continuum mechanics can be applied to the so-called "vacuum" of space. This means that space can be considered as a kind of transparent elastic field. That is, space as a vacuum performs the motion of deformation such as expansion, contraction, elongation, torsion and bending. The latest expanding universe theory (Friedmann, de Sitter, inflationary cosmological model) supports this assumption. We can regard space as an infinite elastic body like rubber.

2) From the General Relativity, the major component of curvature of space (hereinafter referred to as the major component of spatial curvature) R^{00} can be produced by not only mass density but also magnetic field B as follows (see Appendix D):

$$R^{00} = \frac{4\pi G}{\mu_0 c^4} B^2 = 8.2 \times 10^{-38} B^2 \,,\tag{1.43}$$

where $\mu_0 = 4\pi \times 10^{-7} (H/m)$, $c = 3 \times 10^8 (m/s)$, $G = 6.672 \times 10^{-11} (N \cdot m^2 / kg^2)$, B is a magnetic field density with Tesla and R^{00} is a major component of special curvature ($1/m^2$).

Eq.(1.43) indicates that the major component of spatial curvature can be controlled by a magnetic field.

3) If space curves, then an inward normal stress "$-P$" is generated (see Fig. **1**).

This normal stress, *i.e.* surface force serves as a sort of pressure field.

$$-P = N \cdot (2R^{00})^{1/2} = N \cdot (1/R_1 + 1/R_2) \,,\tag{1.44}$$

where N is the line stress, R_1, R_2 are the radius of principal curvature of curved surface.

A large number of curved thin layers form the unidirectional surface force, *i.e.* acceleration field. Accordingly, the spatial curvature R^{00} produces the acceleration field α (see Fig. **1**).

4) From the following linear approximation scheme for the gravitational field equation 1) weak gravitational field, *i.e.* small curvature limit, 2) quasi-static, 3) slow-motion approximation (*i.e.* $v/c \sim 1$), we get the following relation between acceleration of curved space and curvature of space:

$$\alpha^i = \sqrt{-g_{00}} c^2 \int_a^b R^{00}(x^i) dx^i \,,\tag{1.45}$$

where α^i: acceleration (m/s²), g_{00}: time component of metric tensor, a-b: range of curved space region(m), x^i: components of coordinate (i=0,1,2,3), c: velocity of light, R^{00}: major component of spatial curvature.

Eq.(1.45) indicates that the acceleration field α^i is produced in curved space.

5) In the curved space region, the massive body "m (kg)" existing in the acceleration field is subjected to the following force F^i(N), from General Relativity:

$$F^i = m\Gamma^i_{jk} \cdot \frac{dx^j}{d\tau} \cdot \frac{dx^k}{d\tau} = m\sqrt{-g_{00}}\, c^2 \Gamma^i_{jk} u^j u^k = m\alpha^i \, , \qquad (1.46)$$

where u^j, u^k are the four velocity, Γ^i_{jk} is the Riemannian connection coefficient, and τ is the proper time.

Eq.(1.46) yields more simple equation from above-stated linear approximation,

$$F^i = m\sqrt{-g_{00}}\, c^2 \Gamma^i_{00} = m\alpha^i = m\sqrt{-g_{00}}\, c^2 \int_a^b R^{00}(x^i)dx^i \, . \qquad (1.47)$$

Setting $i=3$ (*i.e.* direction of radius of curvature: r), we get Newton's second law

$$F^3 = F = m\alpha = m\sqrt{-g_{00}}\, c^2 \int_a^b R^{00}(r)dr = m\sqrt{-g_{00}}\, c^2 \Gamma^3_{00} \, . \qquad (1.48)$$

The acceleration (α) of curved space and its Riemannian connection coefficient (Γ^3_{00}) are given by

$$\alpha = \sqrt{-g_{00}}\, c^2 \Gamma^3_{00} \, , \qquad \Gamma^3_{00} = \frac{-g_{00,3}}{2g_{33}} \, , \qquad (1.49)$$

where c: velocity of light, g_{00} and g_{33}: component of metric tensor, $g_{00,3}$: $\partial g_{00}/\partial x^3 = \partial g_{00}/\partial r$. We choose the spherical coordinates "$ct=x^0$, $r=x^3, \theta=x^1$, $\varphi=x^2$" in space-time. The acceleration α is represented by the equation both in the differential and in the integral form. Practically, since the metric is usually given, the differential form has been found to be advantageous.

6)   The acceleration of space drive propulsion system is based on the solutions of gravitational field equation, which is derived from Eq.(1.49).

As mentioned above, since the acceleration of space drive propulsion system is currently based on the solution derived from de Sitter solution, there is no need of strong magnetic field.

Propulsion Mechanism of Space Drive

Since the propulsion mechanism used magnetic field in the beginning is easy to understand, we explain it using magnetic field. At present, space drive propulsion does not need the strong magnetic field under the favor of de Sitter solution.

As mentioned above, the principle of space drive propulsion system is summarized in the following.

First of all, it is necessary for the space to be curved. Because the curvature of flat space R^{00} is zero (strictly speaking, only 20 independent components of Riemann curvature tensor R_{pijk} are zero), then the acceleration α becomes zero. Such a curved space is generated not only by mass density but also by magnetic field or electric field. In case that the intensities of the magnetic field B and the electric field E are equal, the value of $(1/2 \cdot \varepsilon_0 E^2)$ is about seventeen figures smaller than the value of $(B^2/2\mu_0)$. As a consequence, the electric field only negligibly contributes to the spatial curvature as compared with the magnetic field. Accordingly, it is effective that the space can be curved by a magnetic field. Since the region of curved space produces the field of acceleration, the massive body existing in this acceleration field, *i.e.* curved space, is moved by thrust in accordance with Newton's second law.

In view of the above described principle of propulsion, the spaceship does not move as long as the magnetic field is static. This is because an action of magnetic field to space is in equilibrium with a reaction from space. It is consequently necessary to shut off the equilibrium state in order to actually move the spaceship. As a continuum, the space has a finite strain rate, *i.e.* the velocity of light. When the magnetic field power source is switched off, it takes a finite interval of time for the curved space to return to flat space. In the mean time, spaceship is independent of curved space. It is therefore possible for the spaceship to proceed ahead receiving the action from the acceleration field. Namely, instantaneously switching off the magnetic field breaks the equilibrium state. Being independent of curved space, spaceship is subjected to the action of field during the finite interval to proceed ahead. In general, a body can not move carrying, or together with, a field that is generated by its body. In other words, the body can not move unless the body is independent of the field.

In a surrounding region of spaceship, a magnetic field as an engine produces a curved space. By switching off the magnetic field, in an instantaneous transition interval which the curved space disappears and returns to flat space, the spaceship is independent of curved space. No interaction is present between curved space and spaceship. Here, the switching on-off the magnetic field implies the following consideration. There exists the seed magnetic field to be compressed in the engine system. Using the magnetic flux - compression technology, we compress the seed magnetic field and produce the spatial curvature induced by compressed strong magnetic field (switching on the magnetic field). The power source of spaceship is consumed in the work of compressing the seed magnetic field. After that, we switch off the magnetic field and this implies the shutting off the power source of the compressing magnetic field. Now referring to Fig. **3**, we describe the propulsion mechanism in detail.

As previously described, the space has a finite strain propagation velocity, *i.e.* strain rate (=velocity of light). Even if the magnetic field is switched off, the curved space reverting to flat space needs a finite time, that is, a length of curved space region divided by strain propagation velocity. The spaceship which exists in the curved space can be propelled by the thrust from the field of acceleration, when the curved space returns to flat space. The spaceship cannot be propelled while the magnetic field is switched on (Fig. **3** (b)). This is because the spaceship produces the field of acceleration by itself and a state of equilibrium is held, that is, the action of magnetic field to space and reaction from space hold equilibrium state. However, when the magnetic field is switched off, the state of equilibrium is broken, and therefore, the spaceship can be propelled by the thrust from the field of acceleration (Fig. **3** (c)), because the spaceship is independent of the field. Accordingly, this propulsion system is essentially defined as a pulse propulsion system.

In order to propel the spaceship, the strained surface of space as shown in Fig. **3** (**a**) is preferable in principle. Namely, a space with an anti-symmetric curvature is preferable, that is, flat space in A' region and curved space region in A region.

Practically, the spaceship can be propelled even by the strained surface of space having symmetric curvature, as shown in Fig. **3** (**b**), (**c**). While the spaceship moves to A' region, the curved space in A' region has already returned to flat space as shown in Fig. **3** (**c**). Therefore, the thrust from the acceleration field of A' region does not act on spaceship. In addition, this thrust is proportional to the mass of a volume element that exists in acceleration field. Although the mass of spaceship exists in the A region, the mass of spaceship does not exist in the A' region, therefore the thrust from the A' region can be disregarded from the outset. It is consequently clear that the A' region does not counteract the thrust which results from the A region. The spaceship can get continuous thrust by repeating the pulse-like ON/OFF change of magnetic field at high frequency.

The propagation velocity of the change from flat space to curved space and the propagation velocity of changing from curved space to flat space are both the same, *i.e.* the velocity of light. This is true for both the A' region and A region. Furthermore, the time interval in which the curved space returns to flat space is the same for both the region A' and A. After being accelerated in the A region, the spaceship proceeds into the A' region. Meanwhile, since the curved space in the A' region returns to flat space, the acceleration in the A' region becomes zero. Further, the spaceship has its mass "*m*" mainly in the A region in the mean time and is subjected to the thrust given by $f_{(A)}=ma$. Conversely, the spaceship has not yet its mass appreciably in the A' region, and is subjected to the thrust given by $f_{(A')}=0$. Therefore, the reverse thrust in the A' region does not exist from the outset. After all, we do not need the anti-symmetric curvature as shown in Fig. **3** (**a**).

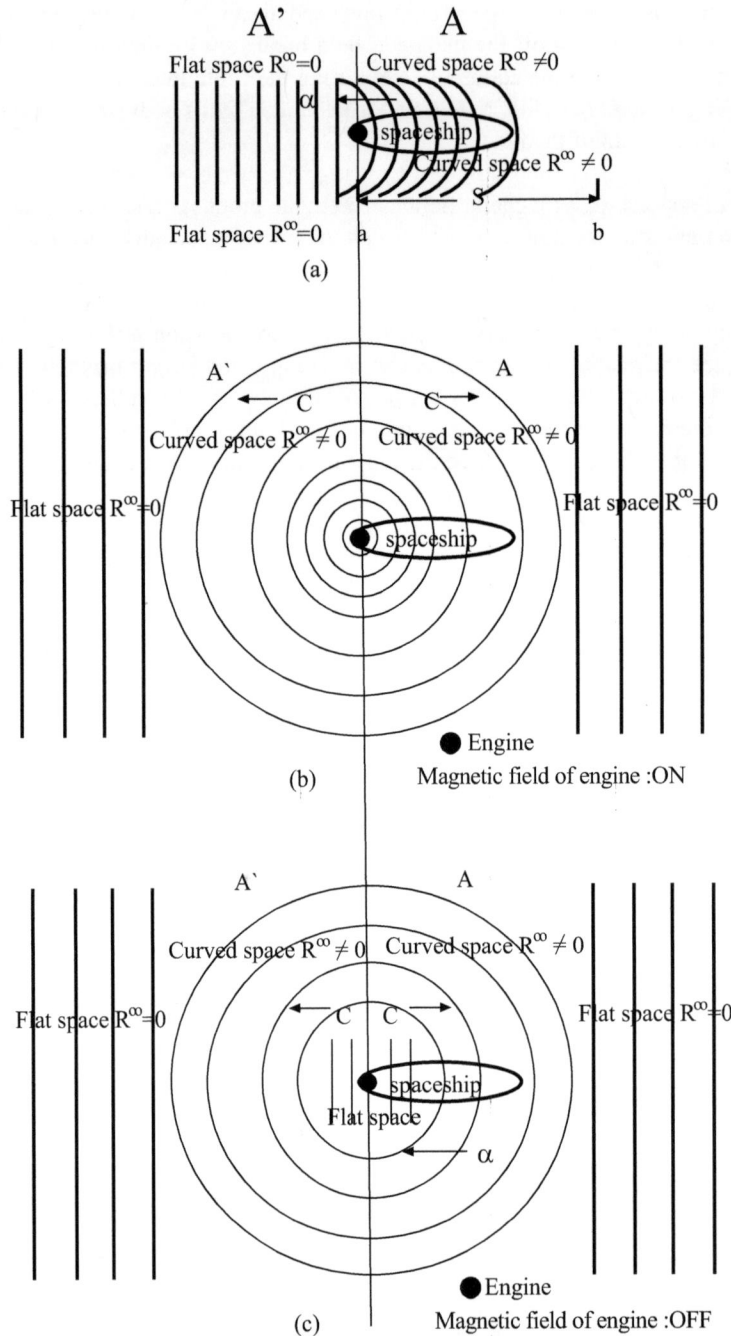

Figure 3: Space drive propulsion principle

It should be noted that the reverse thrust does not occur when the magnetic field is switched off. Some one is bound to think that a switching on and off of an intense magnetic field will make some sort of oscillatory spatial curvature and any net forward motion cannot be imparted to spaceship by this oscillating curvature.

The magnetic field is switched on with a sufficient time in order to produce the curved space region shown by Fig. **3** (**a**). Accordingly, the magnetic field switching off time (t_{OFF}) is much shorter than the magnetic field switching on time (t_{ON}), *i.e.* $t_{OFF} \ll t_{ON}$. Therefore, the curved space region in opposite direction is small enough to be ignored and it cannot produce a reverse acceleration. Because the intensity of acceleration produced in curved space is proportional to both spatial curvature and the size of curved space. Since the region of reverse curvature is very small, the reverse thrust does not occur when the magnetic field is switched off. In addition, we can say that the amount of deviation of curved space in opposite direction is smaller than that of curved space in normal direction. Therefore, the spaceship can be propelled in a single direction.

Some Evaluations of Space Drive Propulsion

Here let us evaluate other features of space drive propulsion such as momentum conservation law, energy conservation law, and the feature of flight performance.

Momentum and Energy Conservation Law

The question is that if the spaceship moves forward, then what moves back? Concerning the space drive propulsion system, the propulsion mechanism is a kind of pressure thrust. As mentioned previously, its propulsion principle is based on the fact that the space is an infinite continuum. We regard the present space as an elastic body described by solid mechanics rather than by fluid dynamics. It may be easy to understand that the spaceship moves by pushing space itself, that is, by being pushed from space. The expression of "moves by pushing space or being pushed from space" indicates that the spaceship produces a curved space region and moves forward by being subjected to the thrust from the acceleration field of curved space. As the motorcar moves by kicking the ground continually infinitely, the spaceship moves by pushing the cosmic space continuously infinitely. The cosmic space as an infinite continuum may be deformed very slightly by being pushed, just like the Earth moves back very slightly by being kicked due to the motorcar. However, this pushing is absorbed by the deformation of space itself continued infinitely. The whole cosmic space is considered as like the ground for kicking. Thus, since the space behaves like the elastic field, the stress between spaceship and space itself is the key of propulsion principle. Accordingly, the analogy of rocket which obeys the momentum conservation law in Newtonian mechanics is not adequate.

If the body (spaceship) in space region gets the energy and the momentum, it means that the outside of body (spaceship), *i.e.* space as a field just loses them. Such a continuity equation means the global physical quantity conservation law. And when the body (spaceship) interacts with the field (space), in order to conserve the energy and momentum as a whole, it is necessary for the field (space) itself to get the energy, momentum, and stress.

Spaceship Flight Performance and Feature

The spaceship equipped with space drive propulsion system has the following features.

a) There is no action of inertial force because the thrust is a body force. Since the body force they produce acts uniformly on every atom inside the spaceship, accelerations of any magnitude can be produced with no strain on the crews, b) The flight patterns such as quickly start from stationary state to all directions in the atmosphere, quickly stop, perpendicular turn, and zigzag turn are possible, c) The final maximum velocity is close to the velocity of light, d) Since the air around the spaceship is also accelerated with spaceship, the aerodynamic heating can be reduced even if the spaceship moves in the atmosphere at high speed (10-100km/s). However, it is expected that a plasma (ionized air) envelops the spaceship, e) Due to the electromagnetic propulsion engine, there is no roar and no exhaust gas, f) The engine and power source are installed in the spaceship, therefore it can fly in the atmosphere of a planet as well as in cosmic space, g) By pulse control of magnetic field, the acceleration varies from 0G to an arbitrary high acceleration (e.g. 36G). h) Deceleration is easy for re-entry into the atmosphere.

Acceleration induced by Cosmological Constant

In the latest cosmology, the terms vacuum energy and cosmological term " Λg^{ij} " are used synonymously. Λ is a constant known as the cosmological constant. The cosmological term is identical to the stress-energy associated with the vacuum energy. The properties of vacuum energy, *i.e.* cosmological term are crucial to expansion of the Universe, that is, to inflationary cosmology. The vacuum energy in de Sitter solution yields the result that the expansion accelerates with time and the total energy with a comoving volume grows exponentially. These facts are due to the elastic nature of vacuum and support the basic concept of space drive propulsion system, that is, the space is an infinite continuum.

Now, from the standpoint of Gauge Theories of the strong, weak, and electromagnetic interactions, there exist the interactions called the "Standard Model" between vacuum and matter (*i.e.* elementary particles). According to the gauge theories, the physical vacuum has various ground states. The potential of vacuum has minima which correspond to the degenerate lowest energy states, either of which may be chosen as the vacuum. Whatever is the choice, however, the symmetry of the theory is spontaneously broken. One of the most important concepts in modern particle theory is that of spontaneous symmetry breaking. The particular interest for cosmology is the theoretical expectation that at high temperatures, symmetries that are spontaneously broken today were restored. During the evolution of the Universe there were phase transitions, perhaps many, associated with the spontaneous breakdown of gauge symmetries. The vacuum structure in many spontaneously broken gauge theories is available for studying field propulsion theory. Accordingly, we can speculate that the above-stated properties of vacuum are preserved even today. As has been stated, the cosmological constant is related to the vacuum energy and phase transition of vacuum.

The most general form of gravitational field equations, which include cosmological constant, is given by

$$R^{ij} - \frac{1}{2} \cdot g^{ij} R = -\frac{8\pi G}{c^4} T^{ij} - \Lambda g^{ij} \,, \tag{1.50}$$

where R^{ij} is the Ricci tensor, R is the scalar curvature, G is the gravitational constant, c is the velocity of light, T^{ij} is the energy momentum tensor, and Λ is the cosmological constant.

Now, concerning the de Sitter cosmological model with non-zero vacuum energy (*i.e.* cosmological constant), the de Sitter line element is written as

$$ds^2 = -(1 - \frac{1}{3}\Lambda r^2)c^2 dt^2 + \frac{1}{1 - \frac{1}{3}\Lambda r^2} dr^2 + r^2(d\theta^2 + \sin^2\theta d\varphi^2) \,. \tag{1.51}$$

The metrics are given by

$$g_{00} = -(1 - 1/3 \cdot \Lambda r^2), \quad g_{11} = g_{22} = 1, \quad g_{33} = 1/(1 - 1/3 \cdot \Lambda r^2),$$
$$and \ \ other \ g_{ij} = 0 \tag{1.52}$$

The acceleration α of de Sitter solution can be obtained finally (see APPENDIX.E)

$$\alpha = \frac{2\pi G\lambda}{3c^2} \phi_0^{\ 4} = 1.6 \times 10^{-27} \lambda \phi_0^{\ 4} \,, \tag{1.53}$$

where G is the gravitational constant, c is the velocity of light, λ is an arbitrary Higgs self-coupling in the Higgs potential (λ is not known and is not determined by a gauge principle, presumably $\lambda \sim 1/10$), and φ_0 is a non-zero vacuum expectation value of Higgs field.

Eq.(1.53) indicates that the vacuum expectation value φ_0 of Higgs field (*i.e.* vacuum scalar field) produces the constant acceleration field. As a result, we find out that the acceleration becomes constant, that is, we can get rid of the tidal force in the spaceship. The scalar field φ can be thought of arising from a source in much the same way as the electromagnetic fields arise from charged particles. We have to search for the fields with the source. The size L of spaceship (*i.e.* length or diameter) is limited to the range r_S, which r_S is the range determined by the following: $V_0(r) = V_0 / r_S \approx 0, \ (L = r_S)$.

Within the range of $L = r_S$, the tidal force in the spaceship and in the vicinity of spaceship can be removed, that is, the acceleration becomes constant within the range of a given region "r_S". The advantageous point of this solution is that even if the size of spaceship is the order of 1km to 10km, the spaceship can move with the constant acceleration (e.g. 3G-30G) to all directions having the flight performance such as quickly start, quickly stop, perpendicular turn, *etc.*

Next, the relation between vacuum potential $V(\varphi)$ and cosmological constant Λ is explained as the following. According to the gauge theories, the physical space as a vacuum is filled with a spin-zero scalar fields, called a Higgs field. The vacuum energy fluctuates in proportion to the fluctuation of Higgs field. The vacuum potential $V(\varphi)$ is given by the vacuum expectation value φ of Higgs field, and we get the minimum of the Higgs potential $V_0(\varphi)$ as follows:

$$V_0(\phi) = \frac{\lambda}{4}\phi_0^{\,4}, \qquad (1.54)$$

where λ is the constant, φ_0 is the non-zero vacuum expectation value of Higgs field.

Since the vacuum potential $V_0(\varphi)$ shall be invariant under the Lorentz transformation, the energy momentum tensor of vacuum $T^{ij}_{\;vac}$ is written in the form as

$$T^{ij}_{\;vac} = V_0(\phi)g^{ij}. \qquad (1.55)$$

The energy momentum tensor of vacuum exerts the same action as that of the cosmological term. It should be noted that $T^{ij}_{\;vac}$ is not energy momentum tensor of matter but the vacuum itself. From Eq.(1.50) and Eq.(1.55), as its metric source, $8\pi G / c^4 \cdot T^{ij}_{\;vac} = 8\pi G / c^4 \cdot V_0(\phi)g^{ij} = \Lambda g^{ij}$, then we get

$$\Lambda = \frac{8\pi G}{c^4}V_0(\phi) = 2.1\times10^{-43}V_0(\phi). \qquad (1.56)$$

Using Eq.(1.53), particular attention is paid to the role of φ_0. Here, only φ_0 is described in NATURAL UNIT ($c = \hbar = k_B = 1$). So, there is one dimension, energy, normally be stated in GeV, that is

$$[\text{Energy}]=[\text{Momentum}]=[\text{Mass}]=[\text{Temperature}]=[\text{Length}]^{-1}=[\text{Time}]^{-1}: \text{in GeV}.$$

In the unit system of natural units "GeV^3" stands for volume density (number density): m^{-3}, and "GeV^4" stands for energy density: $GeV^4 = GeV \times GeV^3 = J/m^3$.

The following relation : $1\text{GeV}^3 = 1.3\times10^{47}\text{m}^{-3}$ is used to convert from natural unit system to SI unit system. The above-stated natural units are used for the field of elementary particle physics or latest cosmology. The vacuum expectation value φ_0 of the present universe is said to be $\varphi_0 \sim 10^{-12}\text{GeV}$ and $\varphi_0^{\,4} = 1\times10^{-46}GeV^4$,

therefore substitution of Eq.(1.54) and Eq.(1.53) with setting $\lambda=1$ gives: $V_0(\varphi) = 1/4 \cdot \varphi_0^{\,4} = 0.5\times10^{-9}J/m^3$, $\alpha = 1.6\times10^{-27}\varphi_0^{\,4} = 3.3\times10^{-36}m/s^2 \approx 0$.

Naturally, the acceleration induced by present cosmic space is zero.

In addition, from Eq.(1.56) and $R^{00} = 4\Lambda$, we get

$$\Lambda = 2.1\times10^{-43}V_0(\varphi) = 1.05\times10^{-52}m^{-2}, \quad R^{00} = 4.2\times10^{-52}m^{-2} \approx 0.$$

Therefore, the present cosmic space is flat space. From $R^{00} = \dfrac{4\pi G}{\mu_0 c^4}B^2$ (see APPENDIX.D), the value of $R^{00} = 4.2\times10^{-52}m^{-2}$ gives the magnetic field of $B = 7.2\times10^{-4}$ gauss (7.2×10^{-8} Tesla). This value of magnetic field agrees quite well with the value of the interstellar magnetic field, *i.e.* $\sim 10^{-5}$ gauss.

If the vacuum expectation value φ_0 of present universe is excited and becomes $\varphi_0 = 6 \times 10^{-3}$ GeV=6MeV, similarly we get the following:

$$V_0(\varphi) = 1/4 \cdot \varphi_0^4 = 6.7 \times 10^{27} J/m^3, \quad \alpha = 1.6 \times 10^{-27} \varphi_0^4 = 43.13 m/s^2 = 4.4G,$$
$$\Lambda = 2.1 \times 10^{-43} V_0(\varphi) = 1.4 \times 10^{-15} m^{-2}, \quad R^{00} = 5.6 \times 10^{-15} m^{-2}.$$

The value of $R^{00} = 5.6 \times 10^{-15} m^{-2}$ gives the equivalent magnetic field of $B = 2.6 \times 10^{11}$ Tesla.

PHASE TRANSITION OF SPACE

The space is a kind of continuum which repeats expansion and contraction. We assume that space as a continuum has two kinds of phases, that is, the elastic solid phase (*i.e.* Crystalline elasticity) like spring and the visco-elastic liquid phase (*i.e.* Rubber elasticity:=Entropy elasticity) like rubber. The elastic solid phase corresponds to the present universe and the visco-elastic liquid phase corresponds to the early universe. Further, we speculate that the space may get the phase transition easily by some trigger, *i.e.* excitation of space, and that the elastic solid phase of space is rapidly transformed to the visco-elastic liquid phase of space and vice versa. The space as a vacuum preserves the properties of phase transition even now. In general, the phase transition is accompanied by a change of symmetry. The phase transition has occurred from an ordered phase to a disordered phase and vice versa.

In a cosmological phase transition, the vacuum expectation value of scalar field φ is transferred from high-temperature, symmetric minimum $\varphi = 0$, to the low- temperature, symmetry-breaking minimum $\varphi = \pm\varphi_0$.

Accordingly, the phase transition is basically related to the spontaneous symmetry breaking, and it is considered that above-stated phenomenon is the fundamental property of space.

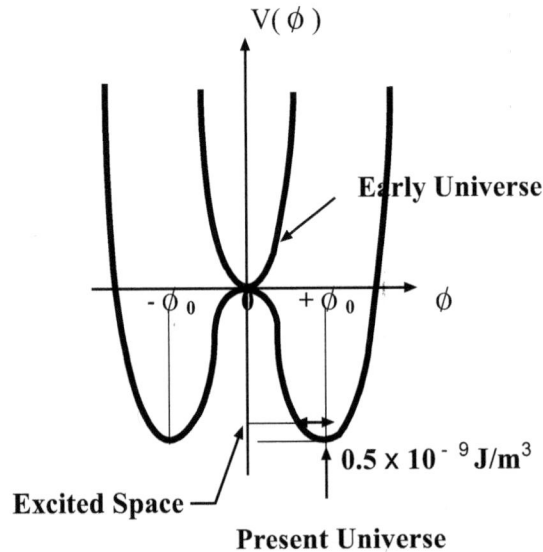

Figure 4: Phase transition of Space

Now, referring to Fig. **4**, the vacuum expectation value of scalar field "$\pm\varphi_0$" indicates the present true vacuum (present universe), and "$\varphi = 0$" indicates the metastable false vacuum in early universe. Even if $\varphi = \pm\varphi_0$ had such a small value, we would expect quantum fluctuations to push φ sufficiently far out on the potential from $\varphi = \pm\varphi_0$ to near the $\varphi = 0$ by a trigger. Since the potential $V(\varphi)$ means the energy density of vacuum corresponding to the value of φ, the value of $V(\varphi)$ directly contributes to the cosmological term. The change in φ gives the change in $V(\varphi)$. As a result, the control of fluctuations of scalar field φ (*i.e.* coherent small oscillations of scalar field) affects the cosmological constant Λ. The enormous vacuum energy of the scalar field then exists in the form of spatially coherent oscillations of the field. As shown in Fig. **4**, a quantum fluctuations to push φ sufficiently by a trigger perturbation of vacuum energy. Therefore, by taking above mechanism used as an unknown technology, we may produce a large cosmological constant, *i.e.* curvature. Since the excitation source of Higgs field (scalar field) is not always restricted to the strong magnetic field, the electric field generated by strong laser focusing effect may be available. Here, the excitation of space means that the value of vacuum expectation value φ is pushed up slightly from its present value $\varphi = +\varphi_0$ and therefore the vacuum potential $V(\varphi)$ is slightly raised.

As conclusion, Fig. **5** shows the summary of space drive propulsion system.

SPACE DRIVE PROPULSION SYSTEM

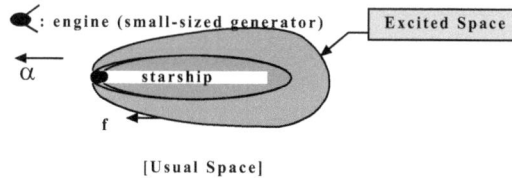

[Usual Space]

Curvature of SPACE (R^∞) plays a significant role for propulsion theory (1988).

$$F^i = m\sqrt{-g_\infty}\,c^2\Gamma^i_\infty = m\alpha^i = m\sqrt{-g_\infty}\,c^2\int_a^b R^\infty(x^i)\,dx^i$$

$$R^\infty = \frac{4\pi G}{\mu_0 c^4}\cdot B^2$$

Acceleration induced by de Sitter solution is found in 1996: constant acceleration α (i.e no tidal force inside of the starship).

$$\alpha = \frac{2\pi G\lambda}{3c^2}\phi_0^4 = 1.6 \times 10^{-27}\,\lambda\,\phi_0^4$$

Φ_n: non-zero vacuum expectation value of field

Figure 5: Summary of Space drive propulsion system

RELATION BETWEEN ALCUBIERRE'S WARP DRIVE AND MINAMI'S SPACE DRIVE

As Gregory L. Matloff stated in his book "DEEP-SPACE PROBES" [24] as follows:

[…Somewhat more immediate are suggestions that we might create an artificial singularity using means other than gravity. Miguel Alcubierre and Yoshinari Minami have independently suggested that we might do this using magnetic field many orders of magnitude greater than those produced on the Earth – even greater than those at the surface of a neutron star or exotic fields that might be manifested from the universal vacuum. Alcubierre's and Minami's ship (if possible) would be pushed or pulled through the Universe by a bubble of warped space-time…], these propulsion theories are well said to be alike.

Further, Edward J. Zampino (NASA Lewis Research Center) states their concept from the viewpoint of energy mainly in his paper entitled "Critical Problems for Interstellar Propulsion System" [42].

In conclusion, both propulsion theories are identical concept from the perspective that they are based on General Relativity and use the idea regarding distortion of space.

However, Alcubierre's warp drive is not manifest for its propulsion principle; there is no mechanism that how local distortion of space-time such as expansion space-time metric or contraction space-time metric create the thrust. Furthermore, Alcubierre's warp drive is a kind of non-used wormholes navigation theory for the purpose of interstellar travel. We added it in chapter.5 from the standpoint of navigation.

interstellar travel. We added it in chapter.5 from the standpoint of navigation.

While, Minami's space drive is manifest for its propulsion principle; there is obvious mechanism that the geometrical structure of space curvature creates actual force as thrust.

Space Coupling Propulsion System

The space propulsion named the space coupling propulsion was proposed Marc G. Millis [32]. Instead of the momentum thrust by the exhaust gas, he claimed that the ZPF field, which is the electromagnetic fluctuation of the vacuum, acts as a reaction mass and the spaceship can move forward by the created artificial gravity.

Four different hypothetical propulsion concepts were proposed [33]. Conservation of momentum requires that the momentum imparted to the vehicle must be equal and opposite to the momentum imparted to a reaction mass. In the case of field drive, there is no obvious reaction mass for the vehicle to push against.

Diametric Drive

The diametric drive considers the possibility of creating a local gradient by juxtaposition of diametrically opposed field sources across the vehicle. This is directly analogous to the negative mass propulsion. Negative mass propulsion is not a new concept. It has already been shown that it is theoretically possible to create a continuously propulsive effect by the juxtaposition of negative and positive mass, and that such a schema does not violate conservation of momentum or energy. A crucial assumption to the success of this concept is that negative mass has negative inertia.

Qualitatively, this concept can be illustrated by the following equation.

$$\varphi_g = (-m)\left[\frac{-G}{\sqrt{(x+d)^2+y^2}}\right] + m\left[\frac{G}{\sqrt{(x-d)^2+y^2}}\right]. \tag{1.57}$$

This gravitational scalar potential is shown as a surface plot over a x-y plane as shown in Fig. **6**.

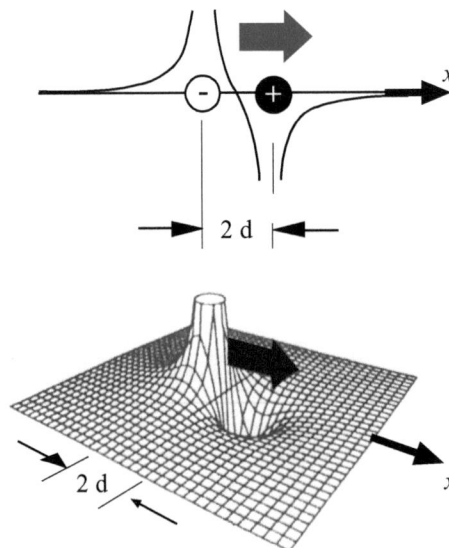

Figure 6: Hypothetical diametric drive

The first term is the gravitational potential for the negative mass, and the second term is for the positive mass. The negative mass and the positive mass are located at a distance d from the center of them. By taking the gradient of the scalar potential caused by the negative mass at the location of the positive mass, and of the positive mass at the location of the negative mass, the acceleration for each mass can be calculated as and

$$a_{+m} = G(+m)/(2d)^2 .$$ (1.59)

Their combined interactions result in a sustained acceleration of both masses in the same direction.

Pitch Drive

The pitch drive (see Fig. **7**) entertains the possibility that somehow a localized slope in scalar potential is induced across the vehicle that causes forces on the vehicle.

In contrast to the diametric drive, it is assumed that such a slope can be created without the presence of a pair of point sources. Qualitatively, this effect can be illustrated by the following equation.

$$\varphi_g = [(-Gm)/r] - xAe^{-r^2} .$$ (1.60)

This scalar potential is shown as a surface plot over the x-y plane as shown in Fig. **7**, which is equal to the superposition of the potential from the vehicle and the induced pitch effect.

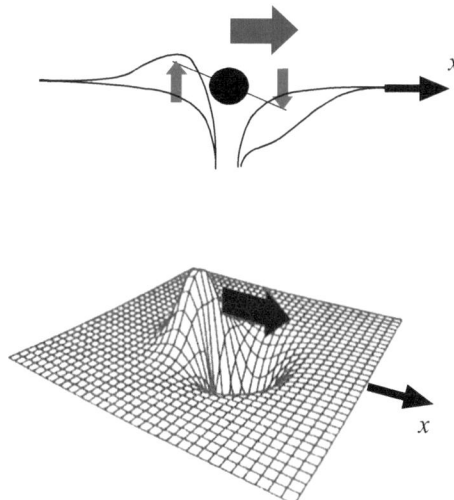

Figure 7: Hypothetical pitch drive

The term on the left is for the vehicle's gravitational potential. The term on the right represents the hypothetical pitch drive effect where a localized gradient in the scalar gravitational potential is induced across the vehicle. This induced pitch effect, which is represented as a magnitude A, with a negative slope in the positive x direction, and is localized by a Gaussian distribution centered at the origin. By taking the gradient of the scalar potential at the location of the vehicle, the acceleration for the vehicle is determined to be A at $r = 0$, and acts in the positive x direction.

Bias Drive

The bias drive (see Fig. **8**) entertains the possibility that the vehicle alters the properties of the space itself, such as the gravitational constant G, to create a local propulsive gradient.

By modifying Newton's constant to have a localized asymmetric bias, a local gradient similar to the pitch drive mechanism results.

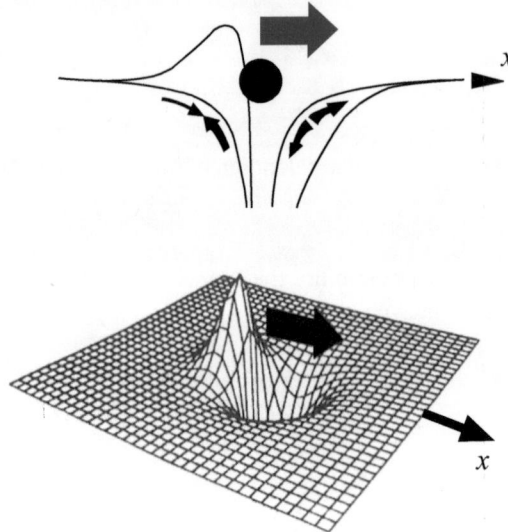

Figure 8: Hypothetical bias drive

Qualitatively, this concept can be given by the following equation.

$$\varphi_g = (xBe^{-r^2} + 1)[-G(m/r)].$$ (1.61)

This scalar potential is shown as a surface plot over the x-y plane in Fig. **8**. This scalar potential includes the Newton's gravitational potential on the right that is multiplied by a spatially asymmetric modifier on the left.

The spatially asymmetric modifier includes a term B, multiplied by x to give a positive slope in the positive x direction, and is localized by a Gaussian distribution as with the pitch drive. A similar concept by Alcubierre suggests creating a propulsive effect by asymmetrically altering space-time itself.

Alcubierre theorized that by using large quantities of negative energy density with an equally large positive energy density, faster-than-light travel would be possible without violating general relativity.

Disjunction Drive

The disjunction drive entertains the possibility that the source of a field and that which reacts to a field can be separated.

By displacing them in space, the reactant is shifted to a point where the field has a slope, thus producing reaction forces between the source and the reactant. It is assumed that the source and the reactant are held apart by some sort of rigid devise.

Qualitatively, this concept can be shown by the following equation.

$$\varphi_g = \frac{-Gm_S}{\sqrt{(x-d)^2 + y^2}}.$$ (1.62)

This scalar potential is shown as a surface plot in Fig. **9**.

It is simply the Newton's gravitational potential of the source mass located at a distance d from the reactant mass. The source mass is defined to have the property that it only causes a field, but does not react to one. The reactant mass is defined to react to the presence of a field, but not to cause one. Thus there is no force on the source mass from the reactant mass.

Supposing that the reactant mass and the source mass are rigidly connected by the device which has pulled them apart, the acceleration becomes

$$a = \left(\frac{G}{d^2} \right) \cdot \left(\frac{m_S m_R}{m_{S_i} + m_{R_i}} \right),$$

(1.63)

where m_{S_i} is a inertial mass of the source mass and m_{R_i} is a inertial mass of the reactant mass.

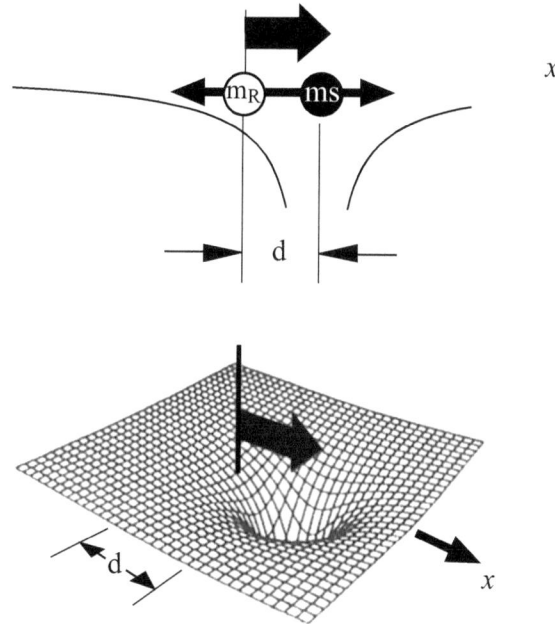

Figure 9: Disjunction drive

There are a variety of unexplored methods toward the physics for space travel.

One of these candidates, it would be useful to seek any means to interact asymmetrically with the medium of electromagnetic fluctuations of the vacuum, which is called ZPF. According to Millis, it is a promising candidates because of its high energy density estimated to be as high as $10^{114} \, J/m^3$. Discovering any way to react asymmetrically with the ZPF would likely create space drive.

Alcubierre's Warp Drive

The warp derive propulsion was proposed by Mexican physicist Alcubierre in 1994.

This is a method to stretch space in a wave, which would cause the fabric of space ahead of spacecraft to contract and the space behind it to expand, as shown in Fig. **10** [1].

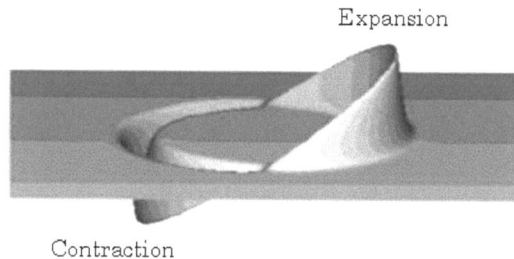

Figure 10: Schematic diagram of the Alcubierre's warp drive

The ship would ride this wave inside a region known as a warp bubble of flat space. Since the ship is not moving within this bubble, but carried along as the region itself moves, conventional relativistic effects such as time dilation do not apply in the way they would in the case of a ship moving at high velocity through flat space-time, even a faster-than-light speed. Also, this method of travel does not actually involve moving faster than light in a local sense, since a light beam within the bubble would still always move faster than the ship; the ship could reach its destination faster than a light beam restricted to traveling outside the warp bubble. Thus, the Alcubierre drive does not contradict the conventional claim that relativity forbids a slower-than-light object to accelerate to faster-than-light speeds. However, there are no known methods to create such a warp bubble in a region that does not already contain one, or to leave the bubble once inside it, so the Alcubierre drive remains a theoretical concept at this time.

The Alcubierre metric, also known as the Alcubierre drive or Warp Drive, is a speculative mathematical model of a space-time exhibiting features reminiscent of the fictional "warp drive" from Star Trek, which can travel "faster than light".

The Alcubierre metric defines the so-called warp drive space-time. This is a Lorentzian manifold which, if interpreted in the context of general relativity, exhibits features reminiscent of the warp drive from Star Trek: a warp bubble appears in previously flat space-time and moves off at effectively superluminal speed.

Inhabitants of the bubble feel no inertial effects. The object within the bubble is not moving (locally) faster than light, instead, the space around them shifts so that the object arrives at its destination faster than light would in normal space

Alcubierre chose a specific form for the function f, but other choices give a simpler space-time exhibiting the desired "warp drive" effects more clearly and simply.

Using the 3+1 formalism of general relativity, the space-time is described by a foliation of space-like hypersurfaces of constant coordinate time t.

The general form of the Alcubierre metric is:

$$ds^2 = -(\alpha^2 - \beta_i \beta^i)dt^2 + 2\beta_i dx^i dt + \gamma_{ij} dx^i dx^j \, , \qquad \qquad \textbf{(1.64)}$$

where α is the lapse function that gives the interval of proper time between nearby hypersurfaces, β_i is the shift vector that relates the spatial coordinate systems on different hypersurfaces and γ_{ij} is a positive definite metric on each of the hypersurfaces. The particular form that Alcubierre studied is defined by:

$$\alpha = 1 \, , \qquad \qquad \textbf{(1.65)}$$

$$\beta^x = -v_s(t)f(r_s(t)) \, , \qquad \qquad \textbf{(1.66)}$$

$$\beta^y = \beta^z = 0 \, , \qquad \qquad \textbf{(1.67)}$$

$$\gamma_{ij} = \delta_{ij} \, , \qquad \qquad \textbf{(1.68)}$$

where $v_s(t) = \dfrac{dx_s(t)}{dt}$, $r_s(t) = \sqrt{(x - x_s(t))^2 + y^2 + z^2}$ and

$$f(r_s) = \frac{\tanh(\sigma(r_s + R)) - \tanh(\sigma(r_s - R))}{2\tanh(\sigma R)} \text{ with } R > 0 \text{ and } \sigma > 0 \text{ which are arbitrary parameters.}$$

Alcubierre's specific form of the metric can thus be written by

$$ds^2 = (v_s(t)^2 f(r_s(t))^2 - 1)dt^2 - 2v_s(t)f(r_s(t))dxdt + dx^2 + dy^2 + dz^2 . \tag{1.69}$$

Two stars are separated with the distance with the distance D in flat space-time and the spaceship is pushed away with a coordinate acceleration that changed rapidly 0 to a constant value a, the time T elapsed in the one way trip becomes

$$T \approx 2\sqrt{\frac{D}{a}} . \tag{1.70}$$

From which, it is clear that T can be made as small as we want by increasing the value of a. Thus the spaceship is able to travel much faster than the speed of light by the propulsion mechanism based on a local distortion of space-time.

With this particular form of the metric, it can be shown that the energy density measured by observers whose 4-velocity is normal to the hypersurfaces is given by

$$\varepsilon = -\frac{c^4}{8\pi G}\frac{v_s(x^2 + y^2)}{4g^2 r_s^2}\left(\frac{df}{dr_s}\right)^2 , \tag{1.71}$$

where g is the determinant of the metric tensor.

Thus, as the energy density is negative, one needs exotic matter to travel faster than the speed of light. The existence of exotic matter is not theoretically ruled out, the Casimir effect and the accelerating universe both lending support to the proposed existence of such matter. However, generating enough exotic matter and sustaining it to perform feats such as faster-than-light travel (and also to keep open the 'throat' of a wormhole) is thought to be impractical. Low has argued that within the context of general relativity, it is impossible to construct a warp drive in the absence of exotic matter. It is generally believed that a consistent theory of quantum gravity will resolve such issues once and for all. For those familiar with the effects of special relativity, such as Lorentz contraction and time dilation, the Alcubierre metric has some apparently peculiar aspects. In particular, Alcubierre has shown that even when the ship is accelerating, it travels on a free-fall geodesic. In other words, a ship using the warp to accelerate and decelerate is always in free fall, and the crew would experience no accelerational g-forces. Enormous tidal forces would be present near the edges of the flat-space volume because of the large space curvature there, but by suitable specification of the metric, these would be made very small within the volume occupied by the ship.

Some Problems of the Warp Drive

In general relativity, one often first specifies a plausible distribution of matter and energy, and then finds the geometry of the space-time associated with it; but it is also possible to run the Einstein field equations in the other direction, first specifying a metric and then finding the energy-momentum tensor associated with it, and this is what Alcubierre did in building his metric. This practice means that the solution can violate various energy conditions and require exotic matter [39]. The need for exotic matter leads to questions about whether it is actually possible to find a way to distribute the matter in an initial space-time which lacks a "warp bubble" in such a way that the bubble will be created at a later time. Although quantum field theory permits the existence of region with negative energy densities in some special circumstances, so this requirement does not immediately eliminate the use of space-time distortion for hyper interstellar travel.

Yet another problem is that, according to Serguei Krasnikov, it would be impossible to generate the bubble without being able to force the exotic matter to move at locally faster-than-light (FTL) speeds, which would require the existence of tachyons. Some methods have been suggested which would avoid the problem of tachyonic motion, but would probably generate a naked singularity at the front of the bubble.

Krasnikov proposed that, if tachyonic matter could not be found or used, then a solution might be to arrange for masses along the path of the vessel to be set in motion in such a way that the required field was produced. But in this case the Alcubierre Drive vessel is not able to go dashing around the galaxy at will. It is only able to travel routes which, like a railroad, have first been equipped with the necessary infrastructure.

Significant problems with the metric of this form stem from the fact that all known warp drive space-times violate various energy conditions. It is true that certain experimentally verified quantum phenomena, such as the Casimir effect, when described in the context of the quantum field theories, lead to stress-energy tensors which also violate the energy conditions and so one might hope that Alcubierre type warp drives could perhaps be physically realized by clever engineering taking advantage of such quantum effects. However, if certain quantum inequalities conjectured by Ford and Roman hold, then the energy requirements for some warp drives may be absurdly gigantic, e.g. the energy equivalent of 10^{67} gram might be required to transport a small spaceship across the Milky Way galaxy. This is orders of magnitude greater than the mass of the universe. Counterarguments to these apparent problems have been offered, but not everyone is convinced they can be overcome.

Reviewing the current status of the warp drive, there are many obstacles to make warp drive foe interstellar travel. Some of major problems are given as follows[21];

(1) Energy violation: Warp drive requires negative energy, both at subliminal and superluminal speeds.

(2) Energy source: Source of negative energy is unknown, which is not allowed by classical physics, but it is not prohibited by laws of quantum physics (Casimir energy).

(3) Blue shift effect: Blue shift of front arriving photons to high energy, creating radiation hazard to vehicle and crew. The warp bubble is surrounded by a bath of radiation, which increases in temperature as the velocity increases. The Unruh effect is a consequence of an accelerating observer in a vacuum.

(4) Causal disconnection; Possible causal disconnection of warp bubble from external negative energy region, so that it could not be controlled or contained. This is a result of the formation of past and future event horizons surrounding the vehicle, when its velocity exceeds the light speed.

(5) Tidal forces; At the boundary between the internal flat space part of the metric and the external flat normal space, enormous tidal force will be present due to the extreme associated space-time warpage there.

They suggest a restriction of warp drive and the difficulty to attain superluminal speed, but the warp drive interstellar travel remains a potential candidate for future generations. However, if vacuum polarization induced interactions between a photon and a gravitational field is taken into consideration, then Drummond and Hathrell have shown that in a Schwarzschild field photons in their one-loop state may propagate faster-than-light, which shown that there exist a possibility for superluminal velocities [6]. Warp drive is a faster-than-light (FTL) propulsion system in the universe of many science fiction settings, most notably including Star Trek. A spacecraft equipped with a warp drive may travel at velocities greater than that of light by many orders of magnitude, whilst circumventing the relativistic problem of time dilation. In contrast to many other fictional FTL technologies, such as a "jump drive" or a "infinite improbability drive", the warp drive does not permit instantaneous travel between two points; instead, warp drive technology creates an artificial "bubble" of normal space-time that surrounds the spacecraft (as opposed to entering a separate realm or dimension like hyperspace). Consequently, spacecraft at warp velocity can continue to interact with objects in normal space. The idea of warping space as a means of propulsion has enjoyed theoretical study by physicists such as Alcubierre, who has designed his own hypothetical drive. However, an approach that may be facilitated by our present level of technological advancement has yet to be proposed.

The ship rests in a bubble of normal space. At present, there is no known way to naturally or artificially establish a separate, finite space-time region or "bubble" as mentioned, such a region is necessary to locally suspend or encapsulate the spacecraft within its view of a "normal" space-time. Concurrently, external from that region, there would exist a "warped" space-time, through which the separate region travels at velocities exceeding c, the speed of light.

Essentially, the spacecraft would remain in the same place, while space-time ahead of the craft would shrink, expanding again behind it.

General Relativistic Gravity Machine

There is a analogical relation between electromagnetic field and gravitational field by research papers by Harris and Braginsky[2,16].

A particle mass m moving in a gravitational and electromagnetic field has the equation of motion as

$$\frac{d^2 x^\mu}{d\tau^2} + \Gamma^\mu_{\alpha\beta} \left(\frac{dx^\alpha}{d\tau} \right) \cdot \left(\frac{dx^\beta}{d\tau} \right) = \frac{e}{mc} F^\mu_\nu \left(\frac{dx^\nu}{d\tau} \right),$$

(1.72)

where e is the charge of a mass, $\Gamma^\mu_{\alpha\beta} = g^{\mu\sigma}(\partial_\alpha g_{\sigma\beta} + \partial_\beta g_{\alpha\sigma} - \partial_\sigma g_{\alpha\beta})/2$ and $F^{\mu\nu}$ is a electromagnetic field tenor.

With the Einstein's field equation of gravitation given by

$$R_{\mu\nu} = -\frac{8\pi G}{c^4} \left(T_{\mu\nu} - \frac{1}{2} g_{\mu\nu} T \right).$$

(1.73)

From which, Harris obtain equations for the case when the particle is slowly moving compared with the light speed and the gravitational field is sufficiently weak that nonlinear terms in Einstein's field equations can be neglected.

$$\nabla \cdot g = -4\pi G \rho,$$

(1.74)

$$\nabla \cdot K = 0,$$

(1.75)

$$\nabla \times g = 0 \ \left(\nabla \times g = -\frac{\partial K}{\partial t} \right),$$

(1.76)

$$\nabla \times K = -\left(\frac{4\pi}{c} \right) \cdot (4G\rho v) + \frac{4}{c} \frac{\partial g}{\partial t},$$

(1.77)

$$F = e\left(E + \frac{v \times B}{c} \right) + m\left(g + \frac{v \times K}{c} \right) + \frac{mv}{2c} \frac{\partial \phi}{\partial t},$$

(1.78)

where G is a gravitational constant, g is a gravitational field, K is a co-gravitational field or gravitomagnetic field, ρ is a mass density and ϕ is a scalar potential of electromagnetic field. This is similar to the magnetic field in the electromagnetic theory.

Analogous to Harris's equations, following equations were also given by Heaviside and Jefimenko [19], those are given by

$$\nabla \cdot g = -4\pi G \rho,$$

(1.79)

$$\nabla \cdot K = 0,$$

(1.80)

$$\nabla \times g = -\frac{\partial K}{\partial t},$$

(1.81)

$$\nabla \times K = -\frac{4\pi G}{c^2} J + \frac{1}{c^2} \frac{\partial g}{\partial t},$$

(1.82)

$$F = m(v \times K).$$ (1.83)

By using analogy to electromagnetism, Forward has shown in his paper that it is possible to built a machine to create a gravitational field by the system of accelerated masses[9,10].

Time varying magnetic field creates a dipole field, the value of the electric field at the center of the torus is

$$E = -\dot{B} = \frac{d}{dt}\frac{\mu N I r^2}{4\pi R^2},$$ (1.84)

where R is the radius of the torus, r is the radius of one of the loops of wire wound around it and N is the total number of turns.

If we replace the wires with pipes carrying a massive liquid, then the known analogy between the electromagnetic and gravitational field can be applied. Then the equivalent gravitational quantities can be given by

$$g = -\dot{K} = -\frac{d}{dt}\left(\frac{\eta N T r^2}{4\pi R^2}\right),$$ (1.85)

where g is the gravitational field generated by the total accelerated mass as shown in Fig. **11**.

In this figure, it has shown the system whose flow is similar to the current flow in a wire-wound ring or torus.

The mass flow through a pipe wound around a torus, which causes a co-gravitational field in a torus. If mass flow is accelerated, the co-gravitational field increases with time and the dipole gravitational field is created. If the pipe is filled with massive liquid and the liquid is moved back and forth in the pipe rapidly enough, then an alternating push-pull gravity field will be generated at the center of the ring. If the machine was big enough and the liquid was dense enough and moving fast enough, we have a gravity catapult that could launch and retrieve spaceships by its gravity repulsion and attraction.

This machine is in the form of a ring of ultradense matter and the flying body is expelled out the other side of the hole with a greatly increased velocity.

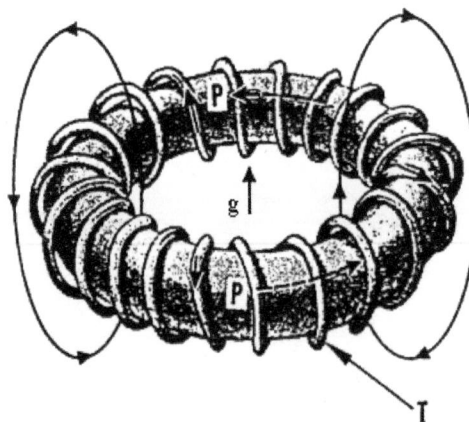

Figure 11: Generator of a Dipole gravitational field

Supposing that the gravitational permeability has the value of $\mu = 3.73 \times 10^{-26} \ m/kg$, we have $g \approx 10^{-10} a$ (a: amount of acceleration) at the center of the torus as shown in Fig. **10**, for the case when the density of a dwarf star through pipes wide as a football field wound around it with kilometer dimensions according to Forward, which is far from present capabilities.

By utilizing this gravitational machine, constant upward gravitational field can be generated which can be used as a gravity catapult that pushes a body as shown in Fig. **12**.

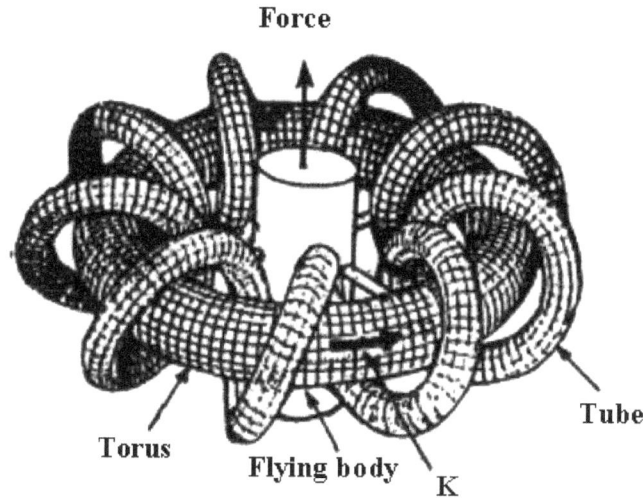

Figure 12: Schematic diagram of the Forward engine

Relativistic Glider

There is a possibility of propulsion to utilize interaction with an external force, which come from the gravitational field of the planet.

If the change of the shape of the system is no longer limited, Gueron presented a purely relativistic effect according to which asymmetric oscillation of a quasi-rigid body slow down or accelerate its fall in a gravitational background[15,41].

First it was presented by Wisdom, who demonstrated that cyclic changes in the shape of a quasi-rigid body may lead to net translation in curved space-times.

This effect was shown to dominate over "swinging" effect for highly enough oscillation frequencies. According to general relativity, test particles follow geodesics on the curved space-time manifold.

The Lagrangian for a particle of mass m is

$$L = mc\sqrt{c^2\left(1-\frac{2GM}{c^2r}\right) - \left(\frac{\dot{r}^2}{\left(1-2GM/c^2r\right)} + r^2(\dot{\theta}^2 + (\sin\theta)\dot{\varphi}^2)\right)}, \qquad (1.86)$$

in the Schwarzschild geometry, the curved space-time around a non-rotating mass, where G is the gravitational constant and c is the light speed.

Fig. **13** shows the example of a quasi-rigid space-time swimmer.

From the vertex, with mass m_0, extend three geodesic struts of proper length l.

In a local Lorentz frame at the vertex, each strut is tilted by the angle α from the axis of the swimmer.

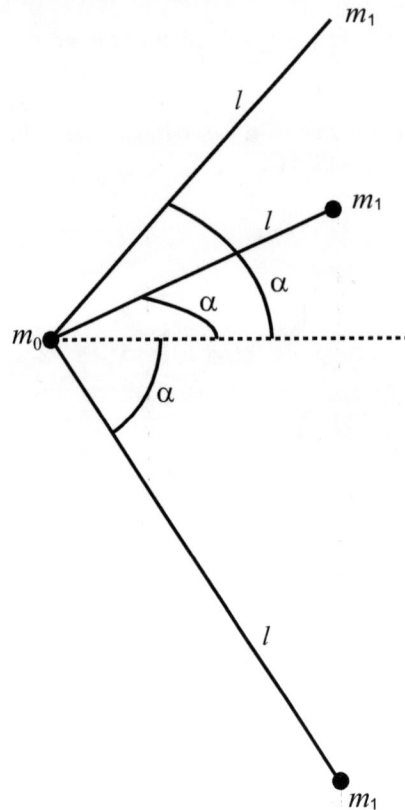

Figure 13: Schematic diagram of the Quasi-rigid swimmer

The three struts are equally spaced around the axis. For Schwarzschild geometry, the components of the Riemann curvature tensor are proportional to GM/c^2r^3, which may be thought of as the inverse of the square of a characteristic radius of curvature. From which, the displacement is proportional to l^2GM/c^2r^3. In addition, the displacement should be proportional to the change in length, to the change in separation angle and to a factor that is homogeneous of degree zero in the masses.

For larger r, where the body is small compared to the radius of curvature of space-time, the displacement cab be given by

$$\Delta\delta = -\frac{3m_0 m_1}{(m_0 + 3m_1)^2} l^2 \frac{GM}{c^2 r^3} \sin\alpha \Delta l \Delta\alpha \; . \tag{1.87}$$

Translation in space can be accomplished merely by cyclic changes in shape, without thrust or external forces. For a meter-sized object performing meter sized deformations at the surface of the Earth, the displacement is of order 10^{-23} m.

Same calculation was conducted for a dumbbell-shaped satellite in near-Earth orbit by M.J.Longo [22]. He estimated the change in the semi-major axis of the orbit after the masses are pulsed together displacement to be

$$\Delta\delta \cong \frac{4}{R} \frac{l^2}{R} \; , \tag{1.88}$$

where R is a radius of the satellite orbit.

For the satellite with $l \approx 1m$, which circulates around near the Earth, the displacement becomes $\Delta\delta \cong 6.3 \times 10^{-7}$ m.

The swimming effect is "geometric" in the sense that it is velocity-independent; the translation due to a certain swimming stroke ("cyclic change in body shape") is entirely determined by the sequence of shapes assumed.

From this effect, a simple system constituted by two masses linked by a massless tether, of which the length changed periodically with time can be achieved for a satellite as a propellantless propulsion, shown in Fig. **14**.

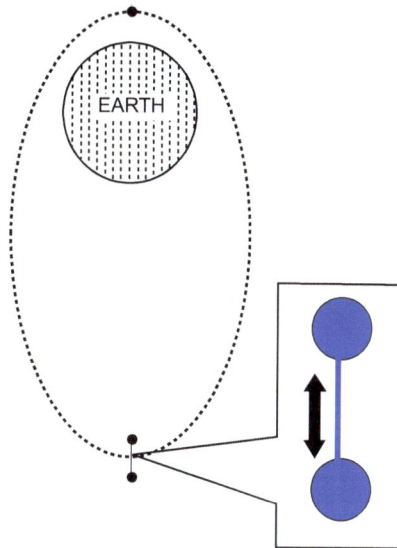

Figure 14: Dumbbell-shaped swimmer.

QUANTUM ELECTRODYNAMICS FIELD PROPULSION

The one of the possible theories for space travel is quantum electrodynamics (QED), which was developed by a number of physicists, beginning in the late 1920s. It deals with the interactions between electrons, positrons and photons and it predicts that the vacuum of space in the universe is filled with electromagnetic waves, random in phase and amplitude and propagating in all possible directions. In this section, we inroduce the basic concept of QED and how it relates to gravity equations of inertia.

Electrogravitic Propulsion System Utilizing ZPF Field

Zero-point Gravitational Theory by Puthoff

Searching to derive Einstein's phenomenological equations for general relativity from more fundamental set of assumptions, Russian physicist Andrei Sakharov (Fig. **15**) came to the conclusion that general relativistic phenomena could be seen as induced effects brought about by changes in the quantum-fluctuation energy of the vacuum due to the presence of matter.

Figure 15 Portrait of Andrei Sakharov

Sakharov suggested that gravity might actually be an induced effect brought about changes in the ZPF in the vacuum, due to the presence of matter. Based on the hypothesis introduced by Sakharov, Puthoff has explored a conceptually simple, classical model of a proposal by Sakharov; namely, that gravitation is not a fundamental interaction at all, but rather an induced effect brought about by changes in the quantum-fluctuation energy of the vacuum when matter is present.

Inertia as formulated by Galileo was simply the property of a material object to either remain at rest or in a uniform motion in the absence of external forces. In his first law of motion, Newton merely restated the Galilean proposition. However, in his second law, Newton expanded the concept of inertia into a fundamental quantitative property of matter. Since that time, there has been only one noteworthy attempt to associate an underlying origin of inertia of an object with something external to that object, and that has been Mach's Principle- the term actually being coined by Einstein. It was argued by Mach that the local property of inertia must somehow asymptotically be a function of the cosmic distribution of all other matter.

Puthoff proposed the hypothesis that ordinary matte is ultimately made of sub-elementary constitutive primary charged entities (parton) bound in the manner of traditional elementary oscillators and he has shown that Lorentz force arises in any accelerated reference frame from the interaction of the partons with the vacuum electromagnetic zero-point field (ZPF)[37].

There appears a ubiquitous ZPF, which can be regarded as a propagating electromagnetic field in free space with spectral energy density shown as

$$\rho(\omega)d\omega = \frac{\hbar\omega^3}{2\pi^2 c^3}d\omega \,,$$ (2.1)

where $\rho(\omega)$ is a energy density of the ZPF field and ω is a radial frequency of the electromagnetic waves.

The spectral distribution in an accelerated frame is given by

$$\rho(\omega) = \frac{\omega^2}{\pi^2 c^3}\left[1+\left(\frac{a}{\omega c}\right)^2\right]\cdot\left[\frac{\hbar\omega}{2}+\frac{\hbar\omega}{e^{\hbar\omega/kT}-1}\right],$$ (2.2)

where a is the proper acceleration relative to a Lorentz frame.

Boyer points out that the additional contribution beyond the thermal (Planck) form is related to the space-time properties of an accelerating reference frame.

$$T = \frac{\hbar a}{2\pi ck} \,,$$ (2.3)

where T is the absolute temp and k is the Boltzmann's constant.

Puthoff et al propose the interpretation that inertia is an electromagnetic resistance arising from the known spectral distortion of the ZPF in accelerated frames. The proposed concept also suggests a physically rigorous version of Mach's principle.

Puthoff found that the inertia of such a particle can also be calculated from the particle's interaction with the ZPF. For the idealized case we have analyzed, the $F = ma$, equation of motion appears to be related to the known distortion of the ZPF spectrum in an accelerated reference frame.

Furthermore, the resistance to acceleration, which defines the inertia of matter appears to be an electromagnetic resistance (specifically Lorentz force) of the ZPF acting at the constituent particle (parton) level. Thus the possibility of developing a scientific version of Mach's principle involving the universal ZPF.

The real issue of whether this field should be regarded as real or virtual has been an ongoing debate in quantum theory.

With respect to the spectral distribution in an accelerated frame given by Puthoff, the force F that the ZPF exerts per constituent parton (elementary sub-particle) in an accelerated frame, it has been found that this force is directly proportional to and directed opposite to the acceleration vector a Newton's law of motion, $F = ma$, may be formulated from the ordinary electrodynamics including the ZPF *via* the techniques of stochastic electrodynamics (SED) given by

$$F = -\frac{1}{2}\Gamma\omega_c \frac{\hbar\omega_c}{c^2}a,$$ (2.4)

where Γ is the Abraham-Lorentz damping constant of the underlying oscillating parton.

From the F-a relationship, the ZPF determined inertia mass associated with the parton oscillator is

$$m_i = \frac{\Gamma\hbar\omega_c^2}{2\pi c^2}.$$ (2.5)

This corresponds to the rest mass associated with the sub-elementary particle, which is a parton. The cutoff frequency comes from the ineffectiveness of the ZPF in producing any translational motion of the parton at wavelength smaller than the parton size. Thus the inertia effect here explored appears primarily because of the distortion of the ZPF vector components at very high frequencies.

Puthoff also shows that the Newtonian gravitational constant G can be determined by the equation:

$$G = \frac{\pi c^5}{\hbar\omega_c^2} = \frac{\pi}{2}c^5 \Big/ \hbar\int_0^{\omega_c} \omega \cdot d\omega,$$ (2.6)

where the integral is taken from zero to ω_c with $\omega_c = \sqrt{\pi c^5 / G\hbar}$.

Furthermore, the spectral energy density diverges as ω_c, with the spectrum effectively cut off at a frequency roughly corresponding to the Planck frequency given by $\omega_p = \sqrt{c^5 / G\hbar}$.

A simple estimate using the value of m_i, the mass becomes

$$m_i = (\Gamma\omega_p)\frac{\hbar\omega_p}{c^2} = \frac{2}{3}\alpha\frac{m_p^2}{m_0},$$ (2.7)

where α is a fine structure constant given by $\alpha = e^2 / \hbar c$ and m_p is a Plank mass given by $m_p = \hbar\omega_p / c^2$. This provided the parton has a charge equal in magnitude to the electron charge.

Under an intense electric field, it has been theoretically predicted that the electron experiences an increase in its rest mass of the form

$$\Delta m = \frac{e^2 <A^2>}{2m_0},$$ (2.8)

where m_0 is the ordinary rest mass of the particle and A is the vector potential.

Let H_A be the electrodynamic Hamiltonian of the particle under high electromagnetic field, it has the form shown as[17]

$$H_A = \frac{e^2}{2m_0c^2} <A^2>,$$ (2.9)

which was analogically discovered by Milonni shown in the paper by Haish, Rueda and Puthoff, where m_0 is the rest mass of the particle, e is its charge and A is the vector potential.

After expanding A in terms of its creation and annihilation operations and averaging in the standard fashion, Haisch, Rueda and Puthoff obtained [17]

$$H_A = \frac{e^2}{2m_0c^2} \sum_{k,\lambda} \left[\frac{2\pi\hbar c^2}{\omega_k V}\right] <a_{k\lambda}a^\dagger_{k\lambda}> = 2\left[\frac{e^2}{2m_0c^2}\right]\left[\frac{2\pi\hbar c^2}{V}\right]\left[\frac{V}{(2\pi)^3}\right]\int \frac{d^3k}{\omega_k} = \frac{e^2\hbar}{2\pi m_0c^3}\omega_c^3.$$ (2.10)

When we consider that a cutoff frequency is a resonant frequency, which is specific to a given particle, the mass of a particle can be converted by external boundary conditions. They suggested that if one could somehow modify the vacuum medium then the mass of a particle or object in it would change according to the zero-point field theory, and this opens up the possibility of manipulation of inertia and gravitation of matter since both properties are shown to stem as least in part from electrodynamics. Thus an inertial drive, which would act exactly like a controllable gravity field, opens up new possibility as one of the possible propulsion system as stated by A.C.Clarke (Fig. **16**) in his "3001, The Final Odyssey".

So called empty space is actually a cauldron of seething energies-the Zero Point Field. HR&P suggest that both inertia and gravitation are electromagnetic phenomena, resulting from interaction with this field. If HR&P's theory can be proved, it opens up the prospect of anti-gravity "space drives", and the even more fantastic possibility of controlling inertia.

Arthur.C.Clarke

Figure 16: Foresight of Future space propulsion by Clarke

Manipulation of Inertia by the Shift of Cut-Off Frequency of ZPF field *via* the Electromagnetic Field

Recent work implies that the ZPF may play an even more significant role as the source of inertia and gravitation of matter [18,38]. Furthermore, this close link between electromagnetism and inertia suggests that it may be fruitful to investigate to what extent the fundamental physical process of electromagnetic radiation by accelerated charged particles could be interpreted as scattering of ambient ZPF radiation.

As the expression for inertial mass derived by Haisch, Rueda and Puthoff (HRP involves two free parameters, Γ and ω_c, which was assumed by HRP some cutoff frequency dictated either by an actual cutoff of the ZPF spectrum or by a minimum size of an elementary particle, as shown in Fig. **17**.

$$\rho(\omega)d\omega = \frac{\hbar\omega^3}{2\pi^2 c^3}d\omega$$

Figure 17: ZPF field in a space

When we consider that a cutoff frequency is a resonant frequency, which is specific to a given particle, the mass of a particle can be converted by external boundary conditions. Haisch, Rueda and Puthoff suggested that if one could somehow modify the vacuum medium then the mass of a particle or object in it would change according to the zero-point field theory.

Under an intense electromagnetic field, it has been theoretically predicted that electron experiences an increase of its rest mass.

Let H_A be the electrodynamic Hamiltonian of the particle under high electromagnetic field, it has the form shown as

$$H_A = \frac{e^2}{2m_0 c^2} < A^2 >, \tag{2.11}$$

which was analogically discovered by Milonni shown in the HRP paper, where m_0 is the rest mass of the particle, e is its charge and A is the vector potential.

The similar equation by using terms of the ZPF field was also obtained by Haisch, Rueda and Puthoff as shown in Eq.(2.10), which can be written by

$$H'_A = \frac{e^2 \hbar}{2\pi m_0 c^3} \omega_c^2, \tag{2.12}$$

where $\hbar$ is a Plank constant divided by 2π and ω_c is a cutoff frequency of ZPF spectrum in the vacuum.

Assuming that electrodynamic Hamiltonians, shown in Eqs. (2.11) and (2.12), are identical with each other, therefore we have $\Delta H_A = \Delta H'_A$ for the dielectric material under impressed electric field as shown in Fig. **18**.

We suppose that the cutoff frequency of the vacuum is shifted as $\omega_c = \omega_0 + \Delta\omega$ when the electromagnetic field is impressed to the dielectric material, $\Delta H'_A$ becomes[35]

$$\Delta H'_A = \frac{e^2 \hbar}{2\pi m_0 c^3}\{(\omega_0 + \Delta\omega)^2 - \omega_0^2\} \approx \frac{e^2 \hbar}{\pi m_0 c^3}\omega_0 \Delta\omega, \tag{2.13}$$

where ω_0 is the Plank frequency given by $\omega_0 = \sqrt{c^5 / \hbar G} \approx 3 \times 10^{43}$ Hz.

No Electric field   **Electric field Impressed**

$A=0$
$H_A = 0$

$H_A = \Delta H'_A$

Initial State $(\Delta H'_A = H'_A - H'_{A0})$

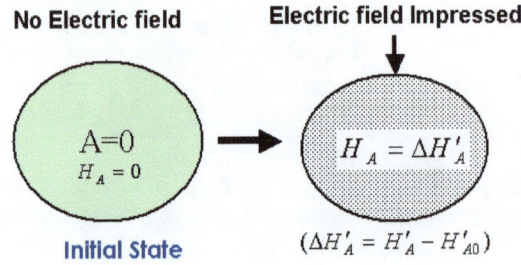

Figure 18: Electrodynamic Hamiltonian with and without electric field

As shown in Fig. **18**, we suppose $H_A = 0$ at the initial state, then we obtain the formula given by

$$\Delta \omega \approx \frac{\pi\, c}{2\hbar\omega_0} < A^2 > .$$

(2.14)

According to the gravitational theory proposed by Haisch, Rueda and Puthoff, we can suppose that the inertial mass of elementary particles induced by ZPF field can be given by

$$m = \frac{\Gamma \hbar \omega_c^2}{2\pi\, c^2},$$

(2.15)

where Γ is the radiation reaction damping constant defining the interaction of charged elementary particles with electromagnetic radiation field. This is a free parameter and it is different from Γ_e which is used for macroscopic electron oscillations in ordinary radiation-matter interactions.

From which, we have

$$\Delta m / m = \frac{2\Delta \omega_c}{\omega_c} = \frac{\pi\, c}{\hbar \omega_0^2} < A^2 > .$$

(2.16)

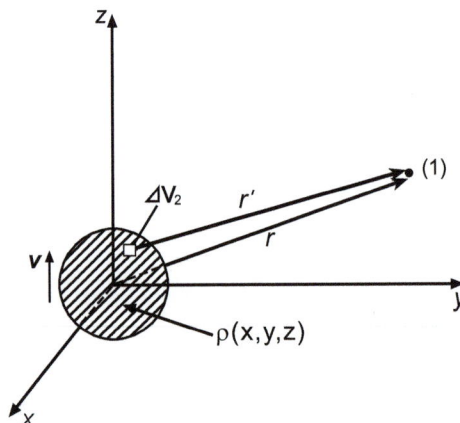

Figure 19: Vector potential created by electric charges

For the dipole field generated by the variance of electric charge as shown in Fig. **19**, the vector potential of the electromagnetic field becomes[7]

$$A = \frac{1}{4\pi\varepsilon_0 c^2} \frac{\dot{p}(t - r/c)}{r} \approx \frac{1}{4\pi\varepsilon_0 c^2} \frac{\dot{p}(t)}{r},$$

(2.17)

where p is a dipole momentum given by $p = qd$ (q : charge of particles, d : displacement of the charge) and ε_0 is a permittivity of free space.

If we let $p(t) = p_0 \sin \omega t$ ($p_0 = Ned$), for the electron cloud as shown in Fig. **20**, we have

$$A = \frac{1}{4\pi\varepsilon_0 c^2} \frac{\omega \; p_0 \cos \omega t}{r} = \frac{1}{4\pi\varepsilon_0 c^2} \frac{\omega Ned \cos \omega t}{r} \; . \tag{2.18}$$

In this equation, N is a number of charges per unit volume and d is given by

$$d = \frac{e}{m} \frac{E}{\omega_e^2 - \omega^2} \; , \tag{2.19}$$

where E is an amplitude of the impressed electric field and ω_e is a resonant angular frequency given by $\omega_e = \sqrt{Ze^2 / \alpha_e m}$ (α_e : electron polarizability), which yields about $10^{15} \sim 10^{16}$ Hz.

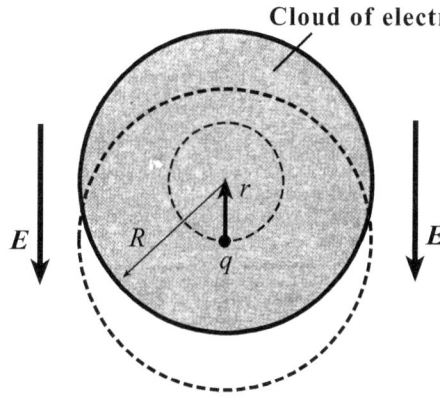

Figure 20: Dipole field generated by the electric field

As the energy dissipation can be incorporated into the analysis by replacing the angular frequency with the complex one given by $\omega' = \omega(1 + i\eta / 2)$, where η is a damping factor which can be given by $\eta \approx \tau_e \cdot \omega$ (τ_e : relaxation time of the electro-dynamical system), we obtain the following equation from above equations for the charged sphere with a radius of R ;

$$
\begin{aligned}
\Delta M(\omega) / M &= \frac{\pi}{c^4} \frac{G}{c^4} \int < A^2 > dv \\
&= \frac{1}{16\pi} \frac{N^2 e^4 G}{\varepsilon_0^2 c^4 m^2} E^2 \left. \frac{\omega^2}{(\omega_e^2 - \omega^2)^2} \right|_{\omega \to \omega'} \int_0^\pi \sin\theta d\theta \int_0^R dr \int_0^{2\pi} d\phi \; . \\
&\approx \frac{N^2 e^4 G}{4\varepsilon_0^2 m^2 c^8} \frac{\omega^2}{(\omega_e^2 - \omega^2)^2 + \eta^2\omega^4} E^2 R
\end{aligned}
\tag{2.20}
$$

From Eq.(2.20), the mass shift of the dielectrics under high potential electric field can be obtained for the cases to impress the alternating electric field and the impulsive electric field, respectively.

Case for the Alternating Electric Field

By impressing alternating electric field to the capacitor composed of dielectric material, Eq.(2.20) can be approximated from

$$\frac{\omega^2}{(\omega_e^2 - \omega^2)^2 + \eta\omega^4} \approx \frac{\omega^2}{\omega_e^4}, \tag{2.21}$$

when satisfying $\omega << \omega_e$.

Then we have

$$\Delta M(\omega) \approx \frac{N^2 e^4 GR}{4\varepsilon_0^2 m^2 c^8} \frac{\omega^2}{\omega_e^4} E^2 M. \tag{2.22}$$

If the high potential electric field shown as $\psi(t) = V_0 \sin \omega t$ is impressed to the dielectrics of the capacitor moving with the displacement given by $x = \delta \sin(\omega_o t)$, where δ is its displacement and ω_o is the oscillating frequency, the new force generated due to the mass shift is given by

$$\hat{F} \approx \dot{x}\frac{d}{dt}M(t) = \gamma \cdot \dot{x} \cdot N^2 R \frac{\omega^2}{\omega_e^4} \frac{M}{d^2}\frac{d}{dt}\psi(t)^2, \tag{2.23}$$

where $\gamma = e^4 G / (4\varepsilon_0^2 m^2 c^8)$.

From which, the amplitude of the generated force becomes

$$\hat{F}_0 = 2\gamma\omega_o \delta \cdot N^2 R \frac{\omega^3}{\omega_e^4} M \frac{V_0^2}{d^2}. \tag{2.24}$$

As the electric power P delivered to the capacitor is proportional to the square of the impressed voltage, the force generated for the capacitor satisfies the relation given by $\hat{F}_0 \propto \delta\omega^3 P$, which is similar to the equation by Mahood on the experiments conducted by Woodward [23].

Case for the Impulsive Electric Field

For the impulsive electric field, which has a wide frequency range of spectrum, the bandwidth of the spectrum ($\omega_2 - \omega_1$) is large compared to the width of the resonance, then the following integration over frequencies across the resonance becomes

$$\int_{\omega_1}^{\omega_2} \frac{\omega^2 d\omega}{(\omega_e^2 - \omega^2)^2 + \eta^2 \omega^4} \approx \frac{\pi}{2\eta\omega_e}. \tag{2.25}$$

From which, we obtain the ratio of the mass shift vs. its rest mass under the impulsive electromagnetic field shown as

$$\Delta M / M = \int_{\omega_1}^{\omega_2} \Delta M(\omega) / M \cdot d\omega = \frac{\pi}{8}\frac{e^4 G}{\varepsilon_0^2 m^2 c^8}\frac{N^2 R}{\eta\omega_e}E^2. \tag{2.26}$$

Assuming that the damping factor η is on the order of the Abraham-Lorenz damping constant given by $\Gamma_e = 2e^2 / 3mc^2$ [17], we have

$$\Delta M \approx \frac{\pi}{8}\frac{e^4 G}{\varepsilon_0^2 m^2 c^8}\frac{N^2 R}{\Gamma_e\omega_e}E^2 M, \tag{2.27}$$

where N is a number of electrons per unit volume in a space including the dielectric material, R is a radius of the electron cloud and E is a magnitude of the impulsive electric field.

Comparing Eqs.(2.22) and (2.27), the ratio of them becomes

$$\Delta M / \Delta M(\omega) \approx 10^{70} / \omega^2 , \qquad (2.28)$$

when satisfying $\omega < \omega_e$. From which, it can be seen that the mass shift by the impulsive electric field to the dielectrics is much greater than the case for impressing alternate electric field, because the impulse electric field has a vast range of spectrum across the resonance, as shown in Fig. **21**.

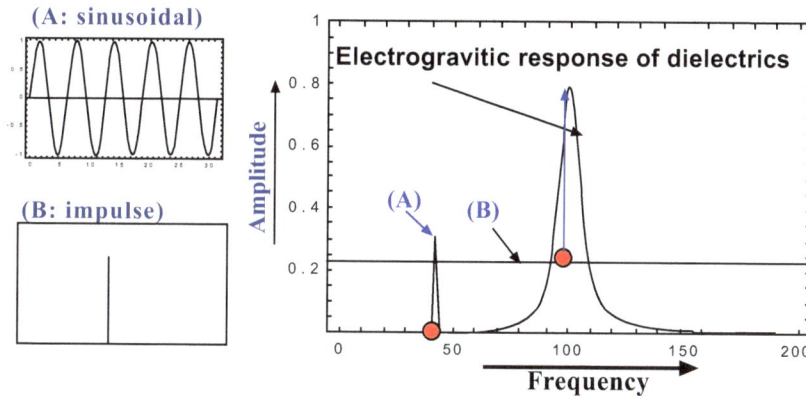

Figure 21: Electrogravitic response of dielectrics both for sinusoidal and impulse waves

Hence it is considered that the impulsive electric field could generate force for applying the space propulsion system.

Modification of Inertia to Generate Propulsion Force

The mechanism to create space drive proposed by Millis has been already given in the previous section. He claimed that the ZPF field acts as a reaction mass and the space ship can move forward by the created artificial gravity. According to his idea, the momentum can be produced for a space drive if the non-symmetrical artificial gravitational field can be created on the spacecraft.

If we suppose that the artificial gravitational field can be created by the external electric field, as shown in Fig. **22**, which gravitational scalar potential according to the formula given in the APPENDIX.I can be written by

$$\varphi_g = -\frac{eE}{m_e}\left(\frac{\delta^2 x}{\delta^2 + (x+d)^2 + y^2} + \frac{\delta^2 x}{\delta^2 + (x-d)^2 + y^2} \right), \qquad (2.29)$$

which is similar to the diametric drive by Millis.

Thus the generated force by an external electric field can be enhanced by coupling with the mass shift and the momentum to propel the spaceship can be produced.

Then the created momentum by the impulsive electric field applied to the spaceship can be obtained as follows;

By using the electrogravitic formula, $E_g \approx -Z\sqrt{4\pi\varepsilon_r\varepsilon_0 G}\cdot E$, which was obtained by the author[34] and Ivanov independently (see Appendix.I), where E_g is a gravitoelectric field, Z is an atomic number of the dielectric material, ε_r is a specific inductive capacity of the dielectrics and E_g is a gravitational field induced by the external electric field, the force produced by electrogravitic field becomes

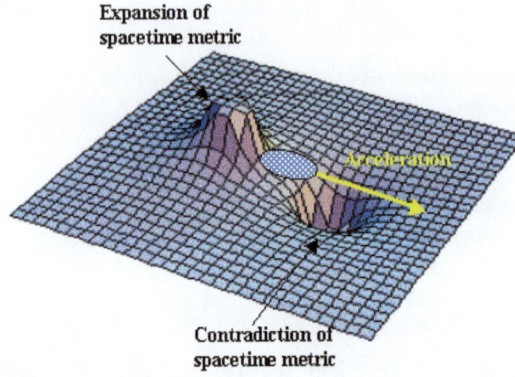

Figure 22: Gravitational scalar potential of the electrogravitic effect

$$F = -(M + \Delta M)E_g \approx Z\sqrt{4\pi\varepsilon \; G}\left(1 + \frac{\pi}{8}\frac{e^4 G}{\varepsilon_0^2 m^2 c^8}\frac{N^2 R}{\Gamma_e \omega_e}E^2\right)EM \; . \tag{2.30}$$

From the equation for the momentum given by $F = dP/dt$, the momentum generated by the gravitational field becomes

$$P_{field} = \int F dt = \int m \cdot E_g dt \approx mE_g \Delta t \; , \tag{2.31}$$

as shown in Fig. **23**.

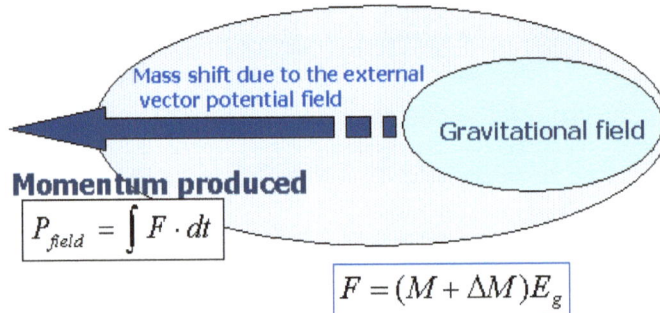

Figure 23: Coupled effect for the ZPF space propulsion

When the electric charge moves between electrodes with a distance l, the time it takes to move that distance becomes $\Delta t = l/v_d$, where v_d is a drift velocity of electrons, the momentum produced by the impulse of charges between electrodes can be given by

$$P_{field} = \int (M + \Delta M) \cdot E_g dt \approx Z\sqrt{4\pi\varepsilon G}M\left(1 + \frac{\pi}{8}\frac{e^4 G}{\varepsilon_0^2 m^2 c^8}\frac{N^2 R}{\Gamma_e \omega_e}E^2\right)\frac{El}{v_d} = Z\sqrt{4\pi\varepsilon G}\left(1 + \frac{3\pi}{16}\frac{e^2 G}{\varepsilon_0^2 mc^6}\frac{N^2 R}{\omega_e}E^2\right)\frac{El}{v_d}M \; . \tag{2.32}$$

From this equation, new factors to induce a momentum for the dielectric material are presented as follows:

- Increase the voltage impressed to the dielectric material, nonlinear increase of the momentum is produced.

- Increase the charge density of electrons in the dielectric material, the greater momentum is produced.

- Increase the separation between the electrodes, the greater momentum is produced.

- Increase the radius of electron clouds, the greater momentum is produced.

Eq.(2.32) suggests that high coronal discharge around the spacecraft which consists of a dielectric material with high charged density of electrons would induce a gravitational force for space propulsion. According to this equation, the greater the radius of the electron cloud, the greater momentum will be produced, hence the space propulsion system for the spacecraft as shown in Fig. **24** can be proposed;

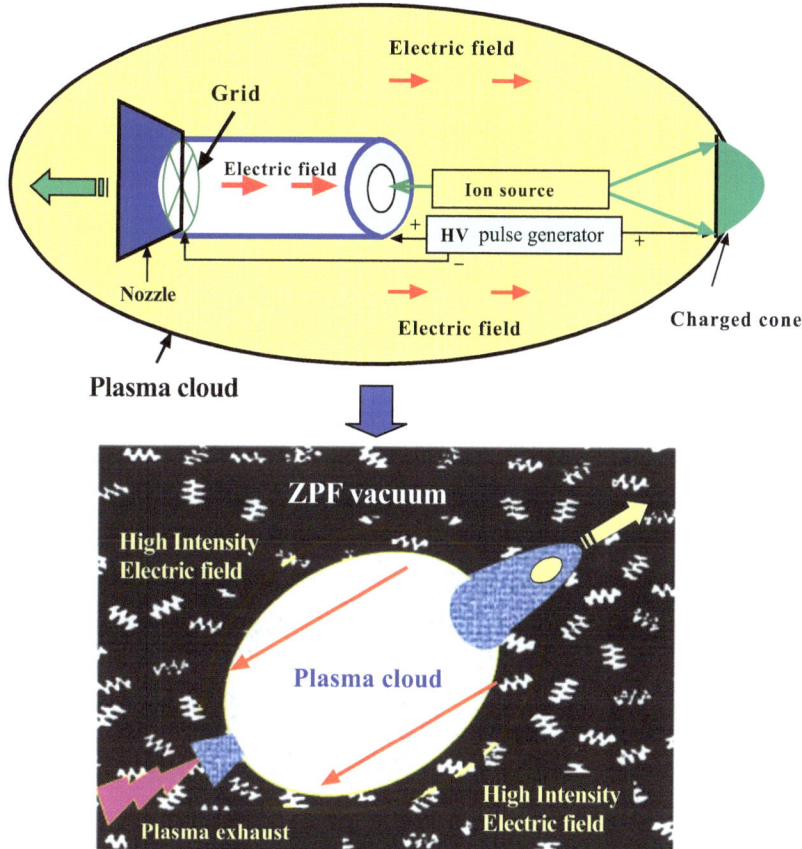

Figure 24: Schematic diagram of the ZPF propulsion system

If we suppose that electromagnetic field generated by the plasma cloud surrounding the spacecraft can influence its mass according to Eq.(2.32), it can attain the higher capability of acceleration by coronal discharges generated by the high-voltage (HV) pulse generator on board as shown in the upper figure of Fig. **24**. This spacecraft is powered by the HV pulse generator for the production of negative ion clouds around the body. The high-voltage power is applied to the cone of the craft and the exhaust nozzle to produce greater potential difference between the cone and the exhaust nozzle. Hence this ZPF propulsion system generates high intensity electric field along its body to produce a locally altered gravity field, which is coupled with the mass shift generated by the coronal discharge that causes a greater forward directed gravitic-force for the spacecraft. The lower figure of Fig. **25** is for the ZPF propulsion mode surrounded by negative ion clouds [36].

From Eq.(2.32), the velocity of the spacecraft generated by the high intensity electric field can be estimated from $v = P_{field} / M'$ (M' : total mass of the spacecraft).

Assuming that the electro-dynamical damping factor has the value on the order of the Abraham-Lorenz damping constant, it can be seen from the calculation result by assuming that $l = 10$ m, $R = 5$ m, $\omega_e = 10^{15}$ rad/s, $m = 9.11 \times 10^{-31}$ kg (electron's mass), $N = 10^{26}$, $v_d = 10^8$ m/s for the value of the vacuum arc[3], and $M \approx M'$, this spacecraft has the possibility to attain the velocity, 9.5×10^5 km/h as shown in Fig. **25**, when applied 10GV

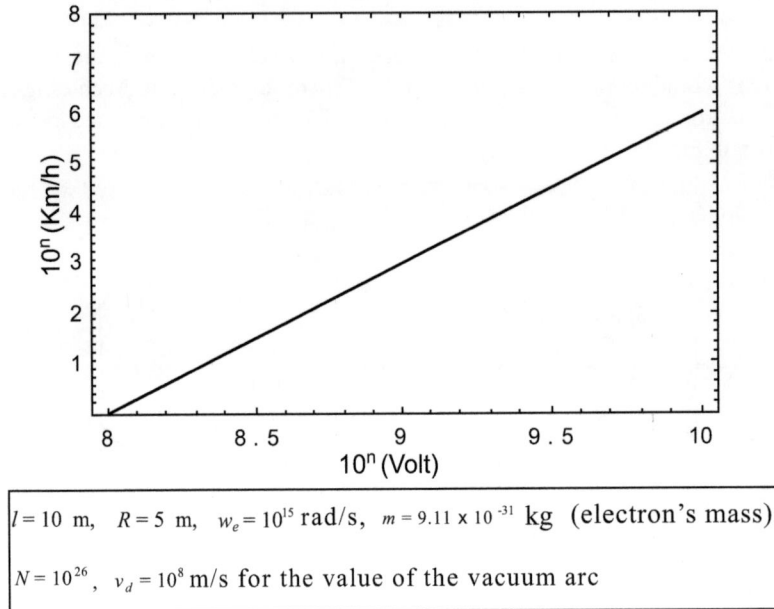

$l = 10$ m, $R = 5$ m, $w_e = 10^{15}$ rad/s, $m = 9.11 \times 10^{-31}$ kg (electron's mass), $N = 10^{26}$, $v_d = 10^8$ m/s for the value of the vacuum arc

Figure 25: Velocity of the Spacecraft by the ZPF propulsion

The electric power of this propulsion system can be estimated from $P = C \cdot \dot{V} V$, where the capacitance of the propulsion system can be obtained from $C = \varepsilon S / l$ (S: area of the capacitor). When we apply alternate current to charge the capacitor given by $V = V_0 \sin \omega t$, the maximum power of the generator yields $P = \omega C V_0^2$ and then we have

$$P_{field} = Z\sqrt{4\pi\varepsilon G}\,\frac{M}{v_d}\sqrt{\frac{P}{\omega C}}\left(1 + \frac{3\pi}{16}\frac{e^2 G N^2 R}{\varepsilon_0^2 m c^6 \omega_e^2 l^2}\frac{P}{\omega C}\right). \tag{2.33}$$

From which, the electric power vs. the velocity of the spacecraft can be estimated as shown in Fig. **26**.

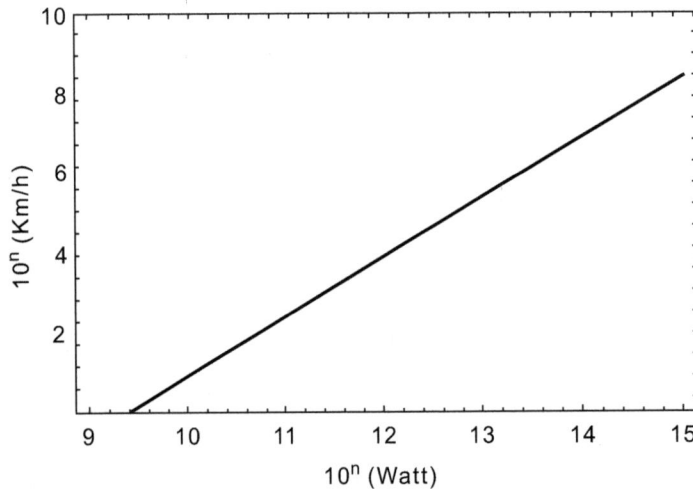

Figure 26: Electric power vs. Velocity of the spacecraft

Froning's Vacuum Propulsion System

The vacuum electromagnetic field and its associated zero-point energies play a key role in the pertubation of electrons, stability of matter and spontaneous emission.

The concept of enginnering the vacuum was first introduced to the physics community by Lee. This new concept is based on the accepted fact that vacuum is charcterized by physical parameters and structure that constitutes an energetic medium which pervades the entire extent of the universe. Space seems inert and empty but quantum field theory and quantum electrodynamics (QED) views it as possessing vigor and vitality over scales of time and space as shown in Fig. **27**.

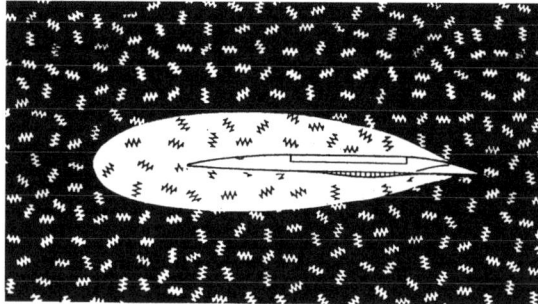

Figure 27: Quantum interstellar ramjet

A major contributor to such vigor and vitality are "zero-point" energy (ZPE) fluctuations, that is, innumerable electromagnetic energy pulsations of varying wavelengths and frequencies which manifest the energetics of the so called "vacuum electromagnetic zero-point field (ZPF)"[12]. Distributions of individual ZPE fluctuations are isotopic throughout undisturbed space and the spectral energy density of ZPF is Lorentz invariant. Thus, the ZPF acts uniformly over bodies moving at a constant speed, causing no net force. Froning adapted a methodology for estimating the relativistic flight performance of a ramjet-like starship.

This method was called a "quantum interstellar ramjet", whose hypothetical propulsion cycle would transform vacuum zero-point energy into the useful radiant kinetic energy of its engine exhaust.

As shown in Fig. **28**, the region of vacuum swept out by the ramjet along the interstellar route is partitioned into sub-regions whose dimentions correspondes to the scale of distance L, over which an expectation value $\hbar c / L$ of zero-point energy could be transformed into a comparable amount of radiant energy within the engine of the ramjet-like ship. If the wave energy emitted by a spaceship in the form of beam interacts with vacuum electromagnetic zero-point field (ZPF), it perturbs ZPF and decreases zero-point radiation pressure, the zero-point radiation pressure in the rear vicinity of spaceship acting upon spaceship. Accordingly, the spaceship moves forward due to the pressure thrust.

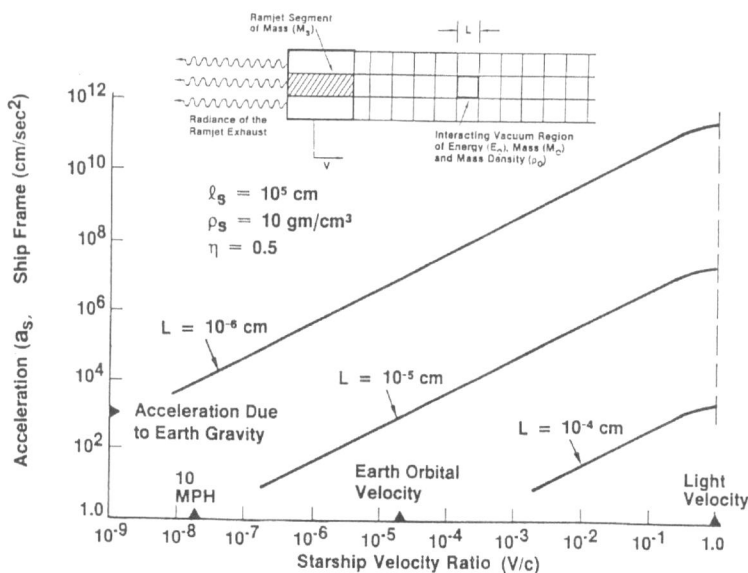

Figure 28: Zero-point radiation pressures in the spaceship vicinity

The accelerating spaceship is bathed in an asymmetric distribution of zero-point radiation pressures, so a pressure thrust will act upon the spaceship. This is because zero-point radiation pressures decrease significantly over the spaceship forward portion for a "pulling" effect and radiation pressures significantly increase over the aft portion of the spaceship for "pushing" effect. The resulting distribution of zero-point radiation pressures acting upon the spaceship result in a thrusting force.

The acceleration of ZPF propulsion system is given as follows:

$$\alpha = [\hbar c^2 / \rho \ell L^4] \cdot [(V / c) / (1 + V / c)] \cdot \eta \,, \tag{2.34}$$

where ρ is a mass density of spaceship, ℓ ia a size of spaceship, V is a velocity, η is a kinetic energy/radiated energy, α is an acceleration, L is the size of perturbation of vacuum, and c is the speed of light.

Expanded Maxwell's Equations for Vacuum Field Propulsion

Solutions to Maxwell's equations represent ordinary electromagnetic radiation which propagates within vacuum in the form of Herzian wave. Barett also observes that such radiation posseses a relative low degree of simmmetry S(1). His expanded version of Maxwell's equations which allow additional phenomenon such as magnetic current density. This gives both a higher SU(2) symmetry by becoming more symmetrical with respect to electric and magnetic terms, given by

$$\frac{\partial \vec{E}}{\partial t} - \nabla \bar{\times} B + iq[\vec{A}_0, \vec{E}] - iq(\vec{A} \times \vec{B} - \vec{B} \times \vec{A}) = -\vec{J} \,, \tag{2.35}$$

$$\nabla \cdot \vec{B} + iq(\vec{A} \cdot \vec{B} - \vec{B} \cdot \vec{A}) = 0 \,, \tag{2.36}$$

$$\nabla \times \vec{E} + \frac{\partial \vec{B}}{\partial t} + iq[\vec{A}_0, \vec{B}] + iq(\vec{A} \times \vec{E} - \vec{E} \times \vec{A}) = 0 \,, \tag{2.37}$$

$$\nabla \cdot \vec{D} - \vec{J}_0 + iq(\vec{A} \cdot \vec{E} - \vec{E} \cdot \vec{A}) = 0 \,, \tag{2.38}$$

where $\vec{J}_0$ is a scalar current.

According to gauge theory, the gauge theory includes both abelian gauge fields like electromagnetic field and non-abelian gauge fields like gravitation. Maxwell's field equations are linear equations, whereas gravitational equations are nonlinear. Since the gravitational field is a non-abelian gauge field and an electromagnetic field is an abelian gauge field, there is not direct interaction between ordinary electromagnetic field and gravitational field. However, if we can transform the ordinary electromagnetic field into a specially constrained electromagnetic field (*i.e.*, a nonabelian field), there is the possibility of its interaction with gravitation.

Froning & Barrett show that such a transformation can increase electromagnetic field symmetry from U(1) to SU(2) and result in non-abelian gauge fields with ability to couple globally with fields of similar form through the action of the vector potential [13]. Thus, if nonabelian gauge field configurations with SU(2) components are associated with gravitation or the electromagnetic zero-point field (ZPF), modification of gravity, inertia, or ZPF might occur within specially conditioned electromagnetic beams (EM beams).

By the way, in particle physics, groups enter because transformations can be carried out on physical systems and the physical systems often are invariant under the transformations. The group U(1) is a one dimensional unitary group. Elements of SU(n) groups are represented by n×n unitary matrices. The set of all 2×2 unitary, unimodular matrices form a group called SU(2). It depends on three continuous parameters.

Fig. **29** shows what must occur within beams of specially conditioned EM radiation emitted from the spaceship in order that the spaceship rapidly accelerates to enormous speeds without excessive stresses or strains exerted upon it. It is seen that such beams must result in compression and expansion of spacetime metric and decreased and increased zero-point radiation pressures in the spaceship vicinity, together with increased em propagation velocity within the beams.

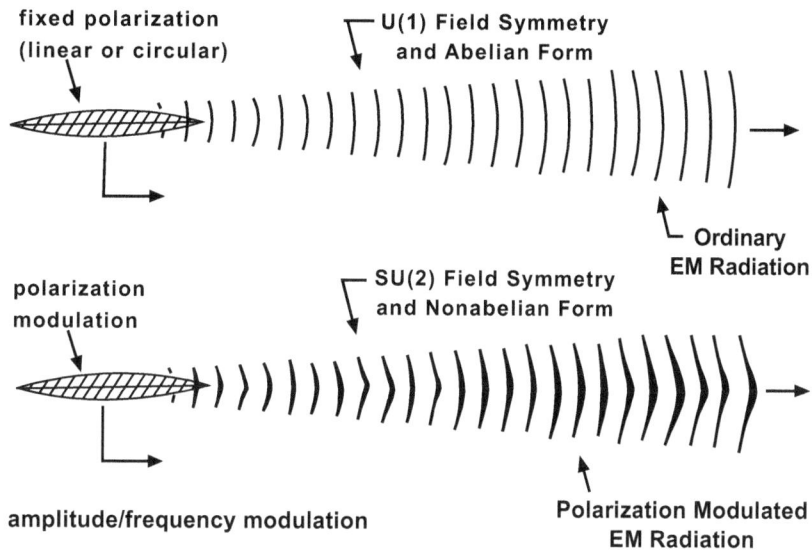

Figure 29: Non-abelian gauge field configurations with SU(2) components, associated with gravitation

The abelian fields associated with ordinary em radiation are U(1) symmetry, while the non-abelian fields associated with weak and strong interactions are of SU(2) and SU(3) symmetry respectively. The more intricate configurations of non-abelian fields result in higher internal symmetries. Gravitation can be described in terms of a non-abelian field. If so, significant interaction between the non-abelian fields associated with gravitation and the abelian fields associated with ordinary em radiation is not likely. This postulates the possibility of such fields coupling globally with the non-Abelian gauge fields that may be associated with spacetime and gravitation through a quantity that may be common to each vector potential, but it is not yet known whether a non-abelian field with higher than U(1) symmetry underlies spacetime metric/gravitation or the ZPF. Though even the most complex combinations of frequency and amplitude modulation do not transform ordinary em fields into non-abelian fields of higher symmetry, such a transformation can be accomplished by modulating the polarization of em wave energy radiated from antennas or apertures of RF or laser transmitters. Here, instead of maintaining a fixed linear or circular polarization, the polarization of the emitted waveform is continually rotated through all possible orientations within time intervals that are extremely short.

Fig. **29** also shows that such polarization modulation increases EM field symmetry from U(1) to SU(2) and results in a non-abelian gauge field with ability to couple globally with fields of similar form through the action of its vector potential. Thus, if nonabelian gauge field configurations with SU(2) components are associated with gravitation, there should be a possibility for modifying gravitational influences within a beam of polarization modulated radiation.

Propulsion System *via* Squeezed Vacuum

A field propulsion system which gets ahead of conventional propulsion systems utilizes the ubiquitous infinite space, more specifically, vacuum. An extraction of thrust from the excited quantum vacuum is indispensable to developing field propulsion.

Gravitation and inertia are an induced effect brought about by changes in the quantum fluctuation energy of the vacuum when matter is present. Gravitation and inertia of matter originate in electromagnetic interactions between the zero-point field (ZPF) and the quarks and electrons constituting atoms. Especially, the attractive gravitational force is akin to the induced van der Waals and Casimir forces [17,18,37]. These play a successful role regarding the fine structure of vacuum. At the present time, although the empirical Casimir forces cannot be directly applied to propulsion, Casimir effects is the sole theory and experiment which indicates the possibility of extracting thrust from vacuum. In order to extract the micro-thrust from vacuum as a first stage, the perturbation of vacuum and the generation of localized inhomogeneous field are indispensable. Here, we define the perturbation of vacuum as the inhomogeneous field of the energy density locally below or above its value in the vacuum state. The vacuum fluctuates with oscillations of the electric field, and these vacuum fluctuations and the zero-point energy have a common origin in the quantized vacuum. The vacuum fluctuations are considered as quantum noise. Until recently, it has been considered that the control of vacuum perturbation was utterly impossible. However, at present, it is proven that the vacuum perturbation can be controlled by squeezed light technology. Therefore, it is possible to increase the energy density locally above the vacuum state and vice versa, decrease the energy density locally below the vacuum state. That is, the squeezed light generates the squeezed vacuum states and yields the coordination geometry of energy density. The theoretical possibility of extracting thrust from the excited vacuum (i.e., squeezed vacuum state) induced by the control of squeezed light and the experimental concepts are proposed by Minami [30,31].

Concerning the squeezed light and squeezed vacuum, their introductions are described in APPENDIX. J.

Vacuum Perturbation Induced by Squeezed Light

The energy density for the squeezed vacuum state was obtained refering to Weigert's article [40]. The following description is based on the cavity which has fixed boundaries, and is focused on the states of the field which minimize the expectation value of the energy in a prescribed region. The spatial variation of the energy density $\hat{W}(x)$ associated with k-th mode is distributed inhomogeneously in the cavity as the following:

$$\left\langle \sigma_k \left| \hat{W}(x) \right| \sigma_k \right\rangle = \frac{\hbar \omega_k}{2LS} \left(\mu_k \sin^2 \omega_k (x + \Lambda) + \frac{1}{\mu_k} \cos^2 \omega_k (x + \Lambda) \right). \tag{2.39}$$

Each mode is a squeezed vacuum state which is denoted by $|\sigma_k\rangle$ since its degree of squeezing is determined by the asymmetry parameter σ_k. Assuming that the Casimir-type cavity, the energy density U near the plate in the state $|\sigma_k\rangle$ (that is, x = λ) is obtained from Eq.(2.39), using $\omega_k = k\pi c/L$, considering a cavity of length L=Λ+λ.

Its boundaries at $x_- = -\Lambda$, $x_+ = \lambda$ are assumed [40]. L is the length of cavity; S is the area of squeezed light.

$$U = \left\langle \sigma_k \left| \hat{W}(\lambda) \right| \sigma_k \right\rangle = \frac{1}{\mu_k} \frac{\hbar \omega_k}{2LS} = \frac{\hbar \omega_k}{2LS} e^{2r} = \frac{1}{\mu_k} \frac{\hbar k\pi c}{2L^2 S} = \frac{\hbar k\pi c}{2L^2 S} e^{2r} \quad (J/m^3). \tag{2.40}$$

By the way, the relation between the squeeze parameter r and μ_k is given by:

$$\mu_k = e^{-2r}. \tag{2.41}$$

On the other hand, the squeeze parameter r is given by (cgs units) [4]:

$$r = \chi \left(\frac{4\pi \omega_s \ell}{cn_s} \right) |E_p| = \chi \left(\frac{4\pi \omega_s \ell}{cn_s} \right) \left(\frac{8\pi P_p}{cn_p A} \right)^{1/2}. \tag{2.42}$$

Here, χ is the effective nonlinear susceptibility, E_p is the amplitude of the pump wave's electric field, ℓ (cm) is the length of the nonlinear medium, n_s and n_p are the index of refraction at the signal and pump frequencies, P_p (erg/s) is the pump power distributed over an area A (cm^2).

From Eq.(2.40), energy density U(z) is the function of squeeze parameter "r". Here, we consider that we control the squeeze parameter in accordance with the position of z axis. Therefore, the squeeze parameter "r" is the function of "z". For simplicity, setting $r(z)=\alpha z$, further, supposing that $\alpha = 1(1/m)$, then $r(z) = z = r$.

Accordingly, we get the force F induced by squeezed light from Eq.(2.40), which generates a vacuum perturbation in the squeezed state as follows:

$$F = -\frac{\partial U(z)}{\partial z} = -\frac{\partial}{\partial z}\left(\frac{\hbar k \pi c}{2L^2 S}e^{2\alpha z}\right) = -\frac{\alpha \hbar k \pi c}{L^2 S}e^{2r} = -\frac{\alpha \hbar k \pi c}{L^2 S}\exp\left(2\chi\left(\frac{4\pi\omega_s \ell}{c n_s}\right)\left(\frac{8\pi P_p}{c n_p A}\right)^{1/2}\right) \quad (N/m^3) . \quad \textbf{(2.43)}$$

Now, let us consider the squeezed vacuum energy density, from another standpoint: Kuo, Ford and Davis showed that the squeezed quantum states generate the negative energy density, in their article Kuo and Ford [20], and Davis [5]. In a squeezed vacuum, $r \neq 0$, $\alpha=0$ (see Eq.(J.3) in APPENDIX.J), also taking $\theta=0$, the expectation value of the stress tensor is given by

$$\langle 0,\zeta | T_{00} | 0,\zeta \rangle = \frac{\hbar\omega}{L^3}\sinh r[\cosh r \cos 2\theta + \sinh r]$$

$$= \frac{\hbar\omega}{L^3}\frac{e^r}{2}\left[\frac{e^r}{2}+\frac{e^r}{2}\right] = \frac{\hbar\omega}{L^3}\frac{e^{2r}}{2} = \frac{\hbar\omega}{2L^3}e^{2r} \qquad , \qquad\qquad \textbf{(2.44)}$$

where

$$\sinh r = \frac{e^r - e^{-r}}{2} \cong \frac{e^r}{2}, \quad \cosh r = \frac{e^r + e^{-r}}{2} \cong \frac{e^r}{2}, \quad \cos 2\theta = 1, \; and \; r \gg 1, \; then \; e^r \gg e^{-r}.$$

Here, T_{00} is the energy density component of stress tensor $T_{\mu\nu}$ in gravitational field.

In the case of large squeeze parameter "r", the energy density accords with described above equation Eq.(2.42), taking notice of $S=L^2$. In addition, considering photon number in squeezed vacuum, the expectation value of photon number is obtained as the following:

$$\langle \hat{n} \rangle = \sinh^2 r \qquad\qquad \textbf{(2.45)}$$

The energy of one photon is $\hbar\omega$, so the energy can be obtained by Eq.(2.46).

$$U_{vac} = \hbar\omega\langle \hat{n} \rangle = \hbar\omega\sinh^2 r \approx \frac{\hbar\omega}{4}e^{2r} \quad (J) . \qquad\qquad \textbf{(2.46)}$$

Then, energy density is obtained by

$$U_{vac} = \frac{\hbar\omega}{4L^3}e^{2r} \quad (J/m^3) . \qquad\qquad \textbf{(2.47)}$$

This accords with described above equation Eq.(2.40) approximately. In any case, a common result is that the vacuum energy density in squeezed state locally increases exponentially as value of squeeze parameter increases in accordance with e^{2r}.

Field Propulsion using Squeezed Vacuum Pressure

The propulsion principle of field propulsion system is pressure thrust induced by a pressure gradient arising from the space-time field (or vacuum field) between the bow and the stern of starship. Since the pressure of the vacuum field in the rear vicinity of starship is high, the starship is pushed through the vacuum field. Pressure of the vacuum field in the front vicinity of the starship is low, so the starship is pulled by the vacuum field. In the front vicinity of starship, the pressure of the vacuum field is not necessarily low but the ordinary vacuum field, that is, just only a high pressure of vacuum field in the rear vicinity of the starship. The pressure gradient of the vacuum field (potential gradient) is formed over the entire range of the starship, so that the starship is propelled by being pushed from the pressure gradient of the vacuum field. To make the starship independent of the pressure gradient of the vacuum field, this propulsion system is essentially defined as a pulse propulsion system. In general, a body can not move carrying, or together with, a field that is generated by its body. In other words, the body can not move unless the body is independent of the field. Therefore, we must keep in mind that propulsion principle is not a kind of "surfing".

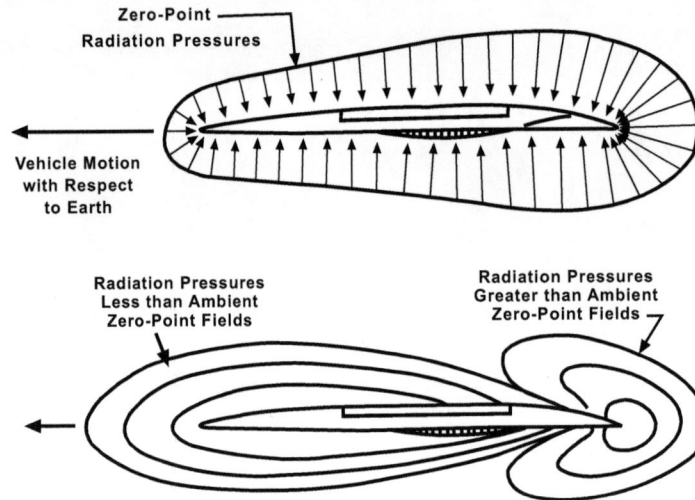

Figure 30: Fundamental propulsion principle of Field propulsion

Referring to the figures below in Fig. **30**, the radiation pressures are less than ambient zero-point fields in the bow of starship, on the while, the radiation pressures are greater than ambient zero-point fields in the stern of starship.

The pressure gradient of the vacuum field is formed over the entire range of starship, so that starship is propelled by being pushed from the pressure gradient of the vacuum field induced by Zero-Point Radiation Pressure [12]. Space-time as a vacuum is generally viewed as a transparent and ubiquitous infinitive empty continuum, upon which physical events take place. However, quantum field theory and quantum electrodynamics (QED) views it as possessing vigor and vitality over scales of time and space. Such vigor and vitality are the zero-point fluctuations of the vacuum electromagnetic field, and the continuous creation and annihilation of virtual particle pairs. We put emphasis on zero-point radiation pressure in this study, that is, potential gradient is generated by gradient of vacuum energy density. Further, vacuum energy density is produced by squeezed light. Zero-Point Radiation Pressure is key factor.

In addition, another view of propulsion principle is described in Fig. **31**.

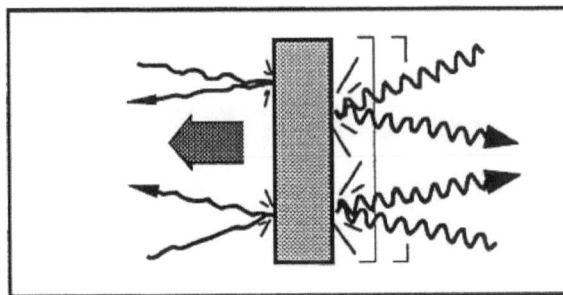

Figure 31: Impulse drive by collision

Referring to Fig. **31**, the collision energy density of the body rear is high, the collision energy density of the body front part is low, and it propels according to the pressure gradient of the body front part and the rear. Analogous to creating a pressure gradient in a fluid, the energy density of the impinging radiation is raised behind the body and lowered in front of the body to create a net difference in radiation pressure across the body.

Next, let us consider propulsion principle again. The propulsion principle described here akin to nuclear propulsion system as shown in Fig. **32**. As is well known, nuclear propulsion is propelled by receiving the action of blast wave of pellet. Similarly, it may be possible to consider that starship is propelled by receiving the action of blast of vacuum perturbation induced by extremely increased vacuum energy density, *i.e.*, strong zero-point radiation pressure. This strong zero-point radiation pressure is periodically generated as impulse drive by pulsed squeezed light.

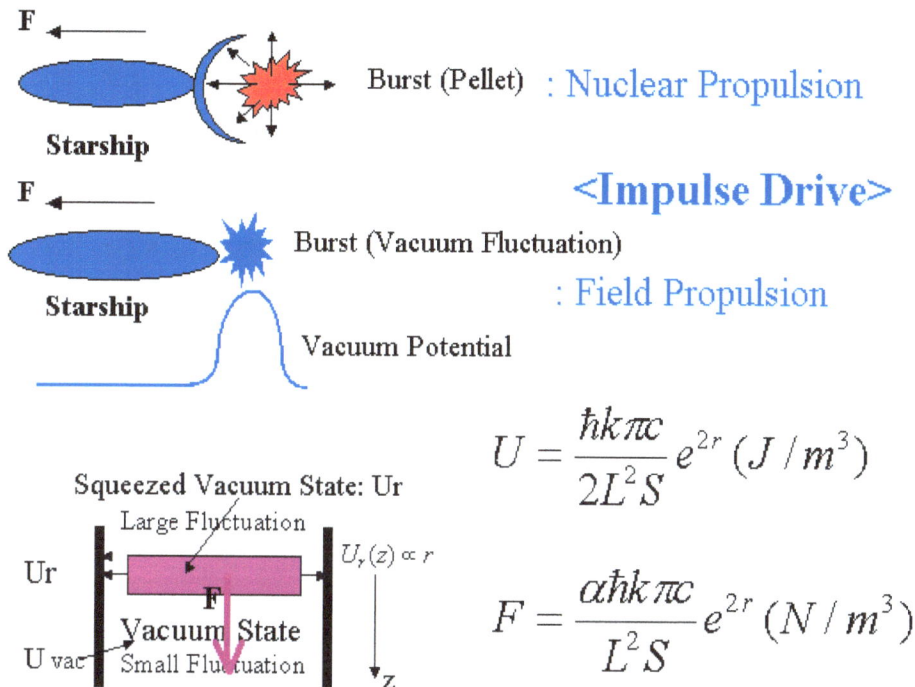

$$U = \frac{\hbar k \pi c}{2L^2 S} e^{2r} \ (J/m^3)$$

$$F = \frac{\alpha \hbar k \pi c}{L^2 S} e^{2r} \ (N/m^3)$$

Figure 32: Propulsion principle & Impulse Drive

As to the above-mentioned propulsion mechanism, a certain background medium exists in each space. This medium has enough large energy and isotropic uniform (isotropy) characteristic features all over space. In order to propel a starship by this medium, a means to control collision power is required. An assumption referred to as that this space has an isotropic medium makes sense. At present, physics admits the Zero-Point Energy (Media of Electromagnetic Fluctuations of the Vacuum: Zero-Point Fluctuations or Zero-Point Field). Space as a vacuum is a kind of actual field, which repeats the creation and annihilation of particle and anti-particle continuously. This state is that the medium of surrounding space itself is considered as reaction mass, and a conservation of momentum law is satisfied.

High-Power Squeezed Laser and Experimental Plan

From here, we would like to show the basic experiment plan. Squeezed state is produced by parametric amplifier. The experimental small chip is connected to hanging wire such as pendulum. When the squeezed vacuum state is instantaneously generated by pulsed squeezed light, the experimental chip may swing receiving the action of blast of vacuum perturbation induced by extremely increased vacuum energy density, *i.e.*, strong zero-point radiation. Fig. **33** shows the basic principle of generating squeezed state by parametric amplifier. A degenerate parametric amplifier consists of a second-order nonlinear crystal pumped by an intense laser beam at angular frequency $\omega_p = 2\omega$. A weak signal beam at angular frequency $\omega_s = \omega$ is also introduced. The nonlinear crystal mixes the signal with the pump and produces an idler beam at angular frequency $\omega_I = \omega$. The pumped nonlinear crystal acts as a phase-sensitive amplifier for signal modes at angular frequency ω.

We assume that there is no signal beam present at the input of the crystal. In this case, the signal is taken from the ever-present vacuum modes. The vacuum modes consist of a randomly fluctuating field of average amplitude. The nonlinear process either amplifies or de-amplifies the vacuum depending on its phase. This produces an output field as shown in squeezed vacuum. That is, with no signal input, the nonlinear crystal amplifies and de-amplifies the vacuum modes, hence producing squeezed vacuum states. As previously mentioned, although the fluctuation in coherent state or vacuum state is temporally constant, the fluctuation in squeezed vacuum state periodically varies from the maximum value of εe^{r} to the minimum value of εe^{-r}. The squeezed light can alter the energy density of the vacuum. The squeezed vacuum exhibits smaller fluctuations, and hence less energy density, than the vacuum in space-time regions. On the other hand, the squeezed vacuum also exhibits larger fluctuations, and hence more energy density, than the vacuum in space-time regions. That is, the oscillatory energy density by periodic occurrences of both smaller and larger fluctuations compared to the unsqueezed vacuum is generated in the cavity. Therefore, the energy density of the vacuum increases exponentially as value of squeeze parameter (r) increases in accordance with "εe^{r}". Inversely, the energy density of the vacuum decreases exponentially as value of squeeze parameter (r) increases in accordance with "εe^{-r}" and is confined within the limits of vacuum value.

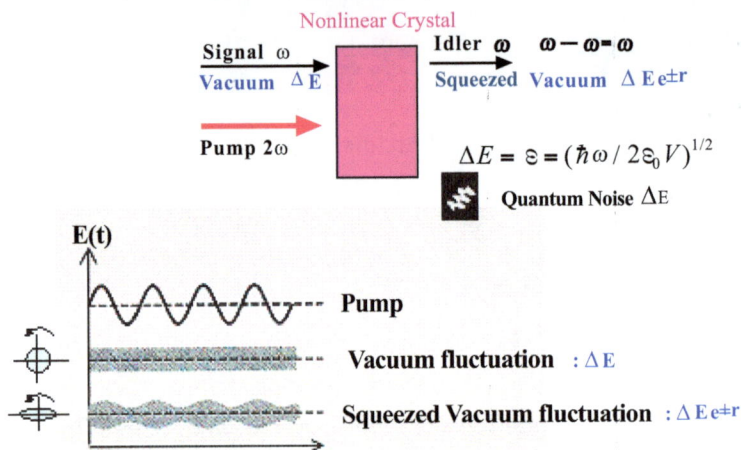

Figure 33: Parametric amplifier

Fig. **34** shows the block diagram to generate the squeezed light for experiment.

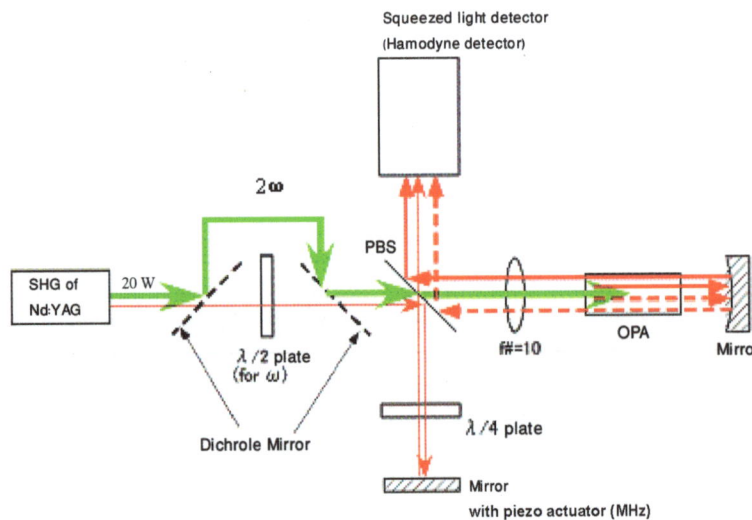

Figure 34: Laser system for high squeezed light

An Nd:YAG laser system with a second harmonic generator (SHG) produces a powerful 20-W laser beam (thick solid line; 2ω). A powerful 20-W laser beam is bypassed by dichroic mirror. The laser beam incident on an Optical Parametric Amplifier (OPA) through beam splitter, PBS (Polarized Beam Splitter for 2nd harmonics) and f# = 10 lens to achieve 10-MW/cm^2 laser intensity inside the OPA for efficient conversion to signal and idler beams, denoted by thin solid and dashed lines. Both beams are reflected by a curved mirror whose curvature and position are designed to output a collimated beam through the f10 lens. A low power fundamental frequency beam from the laser system is redirected by the PBS, reflected by a mirror and incident to squeezed light detector where it serves as a probe beam for homodyne detection of squeezing. The power of the probe is adjusted by a $\lambda/2$ plate. The phase of the probe beam is modulated by a piezo-actuator to see phase dependence of squeezing. Signal and idler beams are also incident on the detector.

Above all, the most important technical future subject is the development of nonlinear crystal having the large value of second-order nonlinear susceptibility for the enhancement of squeeze parameter "r" and so on.

CHAPTER 5

Warp Navigation Methods for Interstellar Travel

Abstract: In order to conquer these huge distances and times for interstellar travel, superluminal speeds become indispensable. If one considers the use of general relativity in the context of interstellar space travel rather than special relativity, then we no longer think of the problem in terms of a propulsion theory, but rather a navigation theory, such as a wormhole engineering of space-time itself. This notion becomes indispensable for the problem if interstellar travel, where cruising ranges are on the order of light-years. Practical interstellar exploration must therefore solve both the propulsion and the navigation problem. In this chapter, several different concepts involving warp drive navigation are described.

INTRODUCTION

Interstellar travel within the span of a human lifetime is impossible as long as we rely on only traditional propulsion systems. Even at the speed of light, the vast distances involved would put most destinations beyond the reach of a single human lifetime. In order to conquer these huge distances and times, superluminal speeds become indispensable. Regrettably the theory of special relativity places restrictions on propulsion theory in that no relative motion can exceed the speed of light, and that there is an asymptotic energy requirement to accelerate to the speed of light that no energy source can overcome.

However, if one considers the use of general relativity in the context of interstellar space travel rather than special relativity, then we no longer think of the problem in terms of a propulsion theory, but rather a navigation theory, in the sense that, in an example such as a wormhole, it is the engineering of spacetime itself that is occurring. This notion becomes indispensable for the problem if interstellar travel where cruising ranges are on the order of light-years. Practical interstellar exploration must therefore solve both the propulsion and the navigation problem. In this chapter, several different concepts involving what we will call warp drive navigation are introduced.

HISTORICAL BACKGROUND

Sixty-three stellar systems and another eight hundred fourteen stellar systems exist respectively within the range of 18 and 50 light years from our Solar system.

For instance, the star Sirius, the seventh nearest star, is 8.7 light years from Earth, while the Pleiades star cluster is 410 light years from us. According to Einstein's Special Relativity, sending a spaceship to a stellar system at a distance longer than several hundred light years would therefore require in excess of those several hundred years even as the spaceship approaches the speed of light. For instance, assuming that the spaceship is traveling to the Pleiades star cluster at a speed of 0.99999c, it will arrive at the Pleiades 1.8 years later and, in case of immediate starting of the return travel, it would be back to Earth 3.6 years after leaving to the Pleiades. But this would be just for the clocks of the astronauts onboard the spaceship for that mission. For people on Earth, the whole time period would be 820 years, with paradoxical consequences as to the feasibility of such a mission.

G. Vulpetti discusses the problems and perspectives of interstellar exploration and shows how and why current physics does not allow real interstellar flight beyond the nearby stars, unless giant world ships are built and the concept of flight through generations is developed: two really formidable tasks indeed [19]. On the other hand G. Vulpetti's "Conscious Life Expansion Principle (CLEP)" that takes interstellar flight to be a fundamental law of Nature would, if true, still compel us to contemplate such as possibility were it our only choice. Further, S. Santoli discusses the possibilities of space exploration that can be envisioned on the basis of the novel emerging technologies that would lead to fully autonomous robots [15].

Therefore some reasonable theoretical speculations are necessary for trying to overcome the limits of the current physics. This was the intent of the recently cancelled NASA Breakthrough Propulsion Physics Program. During the course of this program H.D. Froning showed the theoretical possibility of rapid starship transit to a distant star (*i.e.* Instantaneous Travel) using the method of "jumping" over so-called time and space [7].

Other "spacetime navigation" theories exist that describe spacetime warps, such as Wheeler-Planck wormholes, Kerr metric blackholes, Schwarzschild metric blackholes, and Morris-Thorne field-supported wormhole, and are all based on the solutions of equations of General Relativity [5].

Wormholes come in a variety of species. The major distinction is between Lorentzian and Euclidean wormholes, *i.e.* whether or not the manifold in which the wormhole resides is a Lorentzian (pseudo-Riemannian) manifold, or a true Riemannian manifold with Euclidean signature metrics. Euclidean wormholes are commonly thought of as "instantons" in the gravitational field. Instantons offer the minimum of Euclidean action. Euclidean action implies the operator of imaginary time (*i.e. t→it*). Euclidean wormholes are therefore considered to have a topology with an "imaginary time" coordinate and a nontrivial spatial topology [18].

Let us first consider wormholes as they exist in nature. The size of a typical naturally occurring wormhole is smaller than the atom, *i.e.*, $\approx 10^{-35}$m, and moreover the size is predicted to fluctuate theoretically, making space flight through a naturally occurring wormhole a technical impossibility. Additionally, since the solution of typical wormhole includes a singularity this navigation method would fail in any case due to the loss of spatial integrity of the traveler.

However recent theoretical advances mean that we no longer need be bound by general relativity either. The search for a consistent quantum theory of gravity and the quest for a unification of gravity with other forces (strong, weak, and electromagnetic interactions) have all led to a renewed interest in theories with extra spatial dimensions. Theories that have been formulated with extra dimensions include Kaluza-Klein theory, supergravity theory, superstring theory, M theory, and D-brane theory, which is related to superstring theory. Superstring theory, M theory and D-brane are formulated in 10 or 26 dimensions (6 or 22 extra spatial dimensions). These extra spatial dimensions must be hidden, and are assumed to be unseen because they are compact and small, presumably with typical dimensions of order the Planck length ($\sim 10^{-35}$m) with respect to the traditional spatial dimensions. The navigation method of utilizing extra dimensions for space travel, even if they are compactified, also includes the theoretical possibility of generating a wormhole.

But assuming that space-time is characterized by imaginary time, different Lorentz transformations can then be obtained, and a navigation theory which is based on Special Relativity (not on General Relativity) is proposed as a possibility once again [10,11]. While imaginary time is a difficult concept to grasp, and it is a possibility, and it does not pose any greater difficulty than any of other problems of cosmology. How can the imaginary time have anything to do with the real universe? Stephen Hawking and other cosmologists have been working at developing equations that would describe this to us, and how it is related to what happened just at the instant time began. Though cosmology the concept of imaginary time can be related to the origin and fate of the universe. Hawking's theories use such concepts as imaginary time and singularities to unite relativity and quantum physics [8,9].

WARP DRIVE BY WORMHOLE ENGINEERING

In classical relativity nothing can move faster than the speed of light. Einstein's theories forbid it. In normal space any object approaching the speed of light will increase in mass exponentially, and require an exponential increase in the amount of power needed to propel it forward.

There are two exceptions to this rule however: wormholes, and the warp drive as proposed by Miguel Alcubierre. Let us look first at what is commonly called a wormhole, a bridge connecting two different parts of space, as shown in Fig. **1**.

A ship crossing this bridge would move at below light speed, but yet still arrive before a beam of light that would have gone the long way around.

Space-time geometry of a traversable wormhole can be described by

$$ds^2 = -c^2 dt^2 + dl^2 + (k^2 + l^2)(d\theta^2 + \sin^2\theta \ d\varphi^2) \, , \tag{1.1}$$

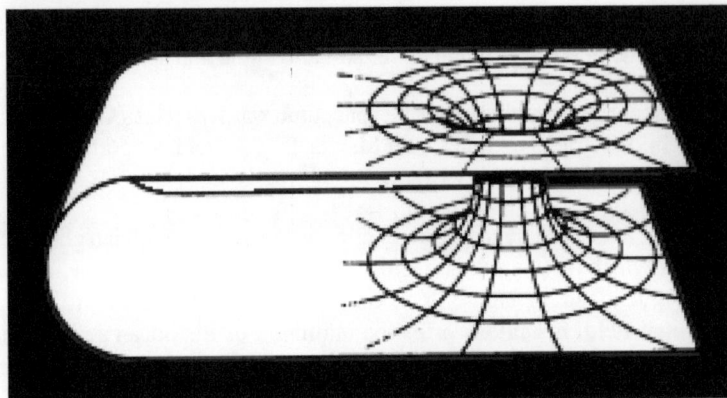

Figure 1: Wormhole connecting two different parts of space

One type of non-traversable wormhole metric is the Schwarzschild solution given by

$$ds^2 = -c^2\left(1 - \frac{2GM}{rc^2}\right)dt^2 + \frac{dr^2}{1 - 2GM/rc^2} + r^2(d\theta^2 + \sin^2\theta\ d\varphi^2),$$

                                                                        (1.2)

The following standard Einstein general relativity field equation permits the travel of an object at a greater velocity than that of light provided that space-time is curved:

$$G_{\mu\nu} = \frac{8\pi}{c^4}GT_{\mu\nu},$$

                                                                        (1.3)

where $G_{\mu\nu}$ is the Einstein curvature tensor, which describes the curvature in space, and $T_{\mu\nu}$ describes the stress-energy tensor, and describes the distribution of mass and energy. They are related through the constant G, which is Newton's gravitational constant. Hypothetically, if space-time is warped properly, the velocity of the traveling object does not need to exceed the speed of light, even though they may appear to be moving faster than light to observers in normal space-time.

Michael S. Morris and Kip S.Thorne provided an overview of wormhole theory in their paper [13]. They have shown that naturally occurring black holes are not traversable because of their enormous tidal forces. The Schwarzschild wormhole for instance could pinch off into infinite tidal forces before anything could possibly get through. However they did outline wormhole solutions to Einstein's field equations that might allow humans to traverse a wormhole. To overcome the problem of traversing an event horizon the wormhole must not have a horizon in its throat. Coupling this constraint with the Einstein field equations leads to a requirement for an exotic form of matter to generate the wormhole's curvature, what could be described as coherent mass, or matter made up of condensed coherent energy. Such coherent matter suppresses the range of naturally occurring zero point motion of the vacuum, and for this reason is often referred to as negative mass or condensed negative energy.

The material must create radial tension of an order approaching the pressure at the center of a massive neutron star. Morris and Thorne observed that for humans to be able to survive a trip through the wormhole, even in a suitable spaceship, this exotic matter must not couple strongly to ordinary matter. By utilizing the exotic matter in a way that compensates for and corrects for tidal forces, travelers in a spaceship may traverse the wormhole throat without encountering any adverse gravitational tidal forces.

To minimize the use of exotic matter, Matt Visser has considered a case in which the wormhole throat has one face that is flat [2]. On such a flat face the radius of curvature would be infinite. A traveler intercepting the flat face would feel no forces and would encounter no exotic matter along this path. Visser reviewed the concept of exotic matter, noting as Morris had that there is negative energy density which is necessary at the throat of the wormhole. He also pointed out that exotic matter (negative matter) is not the same as the antimatter.

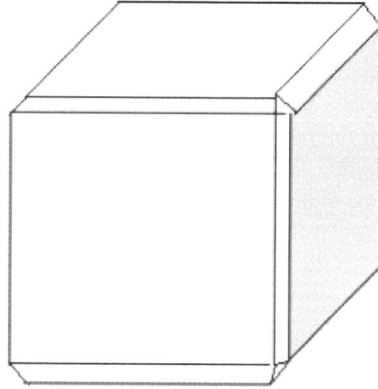

Figure 2: Wormhole throat in the form of a cube with rounded edges and rounded corners[2]

As another example of minimizing the use of exotic matter, Visser considered the possibility that a throat of a wormhole could be constructed of a cube, with rounded edges and rounded corners, which would give six flat faces, all with infinite curvature, with twelve edges that are each constructed of quarter cylinders, and with eight corners which are octants, as shown in Fig. **2**. Given the six faces this exotic construct could be used to connect 6 different pieces of space together. However he estimated that a 1m cube wormhole of such a geometry would require a field with a total negative mass equal in magnitude to that of Jupiter. Clearly this is beyond the reach of current technology.

Thorene and Morris also found a class of exact solutions to the Einstein General Relativity equations that describes field supported wormholes. The metric solution for the Morris-Thorne wormhole is described in [5] by:

As another example of minimizing the use of exotic matter, Visser considered the possibility that a throat of a wormhole could be constructed of a cube, with rounded edges and rounded corners, which would give six flat faces, all with infinite curvature, with twelve edges that are each constructed of quarter cylinders, and with eight corners which are octants, as shown in Fig. **2**. Given the six faces this exotic construct could be used to connect 6 different pieces of space together. However he estimated that a 1m cube wormhole of such a geometry would require a field with a total negative mass equal in magnitude to that of Jupiter. Clearly this is beyond the reach of current technology.

Thorene and Morris also found a class of exact solutions to the Einstein General Relativity equations that describes field supported wormholes. The metric solution for the Morris-Thorne wormhole is described in [5] by:

$$ds^2 = -c^2 dt^2 + \left[\frac{1}{1 - b(r)/r} \right] dr^2 + r^2 d\theta^2 + r^2 \sin^2 \theta \ d\varphi^2 , \qquad (1.4)$$

$$ G_{\mu\nu} = \begin{pmatrix} 1 & 0 & 0 & 0 \\ 0 & \left[\dfrac{1}{1 - b(r)/r} \right] & 0 & 0 \\ 0 & 0 & 1 & 0 \\ 0 & 0 & 0 & 1 \end{pmatrix} , \qquad (1.5)$$

where $b(r)$ is the strength of the exotic field, which is a function of the radial parameter r.

Thorne and his students have designed a space warp that consists of two perfectly conducting spheres completely covering the spherical holes in space-time that form the "mouths" of the wormhole. Although the two spheres are quite large and have a comparatively small door in them to let a spacecraft pass though, their inner surfaces are only a nominal, barely macroscopic distance apart While the distance as measured through the wormhole is very small, the distance between the mouths in the externally observed universe can be light years apart. The two spheres have a large amount of identical electric charge on them. This has two effects. First, the identical charges on each sphere causes them to electrically repel each other. This electrical repulsive force is sufficient to exactly counterbalance the vacuum fluctuations forces trying to pull the two closely spaced conducting spheres together, minimizing the need for exotic matter. Second, the strong electric fields on the outside of the spheres act to keep the wormhole throat open in the region from the outside surface of the sphere off into flat space at infinity.

Richard Obousy proposed an entirely different warp drive design, which could be created by directly manipulating the extra dimensions of string theory:

$$\langle E_{vac} \rangle = -\frac{\pi^2}{R^4}\left[\frac{(2+n)(3+n)}{2} - 1\right][\zeta(0)]^{n-1}\zeta'(4), \tag{1.6}$$

where $\zeta(n)$ is a Riemann zeta function.

In this case the radius of the extra dimensions directly controls the expansion of space. Advanced technology might influence a string and locally adjust the size of the extra dimension, thereby creating a controlled expansion and contraction of the space surrounding the spaceship.

ALCUBIERRE'S WARP DRIVE NAVIGATION

Alcubierre proposed the warp drive propulsion in 1994 [1]. This is a method to stretch space in a wave, which would cause the fabric of space ahead of spacecraft to contract and the space behind it to expand, as shown in Fig. **3**.

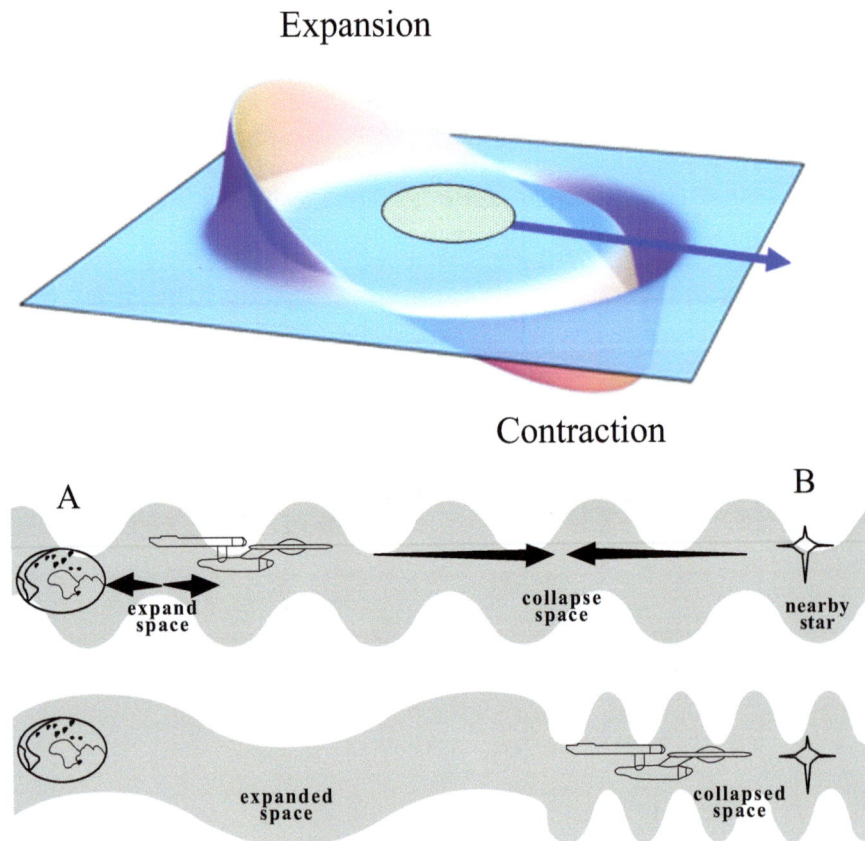

Figure 3: Alcubierre's warp drive concept

The ship would ride this wave inside a region known as a warp bubble of flat space. Since this ship is not moving within this bubble, but is carried along as the region itself moves, it can move at high velocity through flat space, even at a faster-than-light speed. As shown in Fig. 3, two stars A (Earth) and B (a nearby star) are separated by a distance D in flat space-time. At time t_o, consider the case where a spaceship starts to move away from A at a speed v<c (light speed) using its rocket engines. The spaceship then stops at a distance d (d<D) away from A. It is at this point that a disturbance of space-time centered at the spaceship's position, would first be activated.

This disturbance is such that the spaceship is pushed away from A with a coordinate acceleration that changes rapidly from 0 to a constant value a. Since the spaceship is initially at rest v_s=0, the disturbance will develop smoothly from flat space.

When the spaceship is halfway between A and B, the disturbance could be modified in such a way as to effect a coordinate acceleration change rapidly from *a* to *−a*. If the coordinate acceleration in the second part of the trip is arranged in such a way as to be the opposite to the one we had in the first part, then spaceship will eventually find itself at rest with a distance d away from B, at which time the disturbance of space-time could be deactivated, (since again v_s=0). The journey would then be completed by moving again through flat space-time at a sub-light speed *v*. According to Fig. 3, the metric change regarding distortion is not local but global in space-time.

HYPER-SPACE NAVIGATION THROUGH THE "IMAGINARY TIME HOLE"

An interstellar travel within a human lifetime is considered as utterly impossible. The interstellar travel used Special Relativity is well known. However, there exists the problem of the twin or time paradox. In addition, Space warp using wormhole is also well known. However, the size of the wormhole is smaller than the atom, and moreover, the wormhole is predicted to fluctuate theoretically due to instabilities. Assuming hyperspace as being characterized by imaginary time, it is shown that the limitations due to the extremely long time required for interstellar travel as is set forth by Special Relativity are removed. Minami proposed the hyper-space navigation theory used imaginary time in 1993 [10]. This theory is not based on General Relativity but Special Relativity.

As is well known in astronomy, sixty-three stellar systems and other eight hundred fourteen stellar systems exist respectively within the range of 18 and 50 light years from our Solar System. For instance, Alpha Centauri is the nearest star from Earth, and the star Sirius, which is the seventh nearest star, is 8.7 light years from Earth, while the Pleiades star cluster is 410 light years from us. According to Einstein's Special Relativity, sending a starship to a stellar system at a distance longer than several hundred light years would ask for an extremely long time even if the starship would travel at the speed of light. For instance, assuming that the starship is traveling to the Pleiades star cluster at a speed of $0.99999c$, it will arrive at the Pleiades 1.8 years later and, in the case of immediately starting of the return travel, it would be back to Earth 3.6 years after leaving for the Pleiades. But this would be just for the clocks of the astronauts onboard the starship for that mission. For people on Earth, the whole time period would be 820 years, with paradoxical consequences as to the feasibility of a mission such as this. The first solution of the above-stated problem is to obtain a breakthrough in propulsion science. However, no propulsion theory currently exceeds the speed of light. Accordingly, the propulsion theory alone is not enough to establish the reality of interstellar travel, thereby requiring a navigation theory as a secondary solution.

Concerning interstellar travel, the method using a wormhole is well known; relying on space warps, such as for instance Wheeler-Planck Wormholes, Kerr metric, Schwartzschild metric, Morris-Thorne Field-Supported Wormhole based on the solutions of equations of General Relativity [5]. However, since the size of wormhole is smaller than the atom, *i.e.*, $\sim 10^{-35}$m and moreover the size is predicted to fluctuate theoretically due to instabilities, space flight through the wormhole is difficult technically and it is unknown where to go and how to return. Additionally, since the solution of wormhole includes a singularity, this navigation method theoretically includes fundamental problems. There exists another interstellar navigation theory. Froning shows the rapid starship transit to a distant star (*i.e.* Instantaneous Travel) using the method of "jumping" over so-called time and space [7]. Especially, Froning is the first person who applied imaginary time to interstellar navigation. In addition to this invaluable concept, Hyper-Space navigation theory using a space-time featuring an imaginary time offers great promise to develop practical interstellar exploration. This proposed navigation theory is based on Special Relativity (not on General Relativity), that allows interstellar travel to the farthest star systems to be realized; and removes the present theoretical limitations to interstellar travel that arises from the extremely long time needed (the time paradox) according to Special Relativity [10,11,12]. The practical interstellar travel combines propulsion theory with navigation theory.

Three Ways to the Interstellar Travel

Three methods are considered to reach the star rapidly. The basic principle is the following equation which is known to every one:

$$L_{star} = V_{starship} \times t$$

where, L_{star} is the distance to star, $V_{starship}$ is the speed of starship, t is the time.

The distance to a stellar system "L_{star}" is enormous. An extremely long time is required, even if the starship would travel at the speed of light "c". To reach the star rapidly, three parameters, such as "speed", "distance" and "time", shall be controlled.

1) <Change speed> $L_{star} = (nc) \times t$,

where, "nc" is n-fold increase in speed of light "c". Here, n is real number greater than "1".

There is no propulsion theory exceeds the speed of light, moreover, Special Relativity restricts the maximum speed to the speed of light; therefore this method is impossible.

2) <Change distance> $L_{star} / n = c \times t$.

The so-called "wormhole" is utilized. By using wormhole, shorten the distance as $L_{star}/n \approx$ a few meters, as shown in Fig. **4**. For example, one meter in a wormhole corresponds to a few light years in actual space.

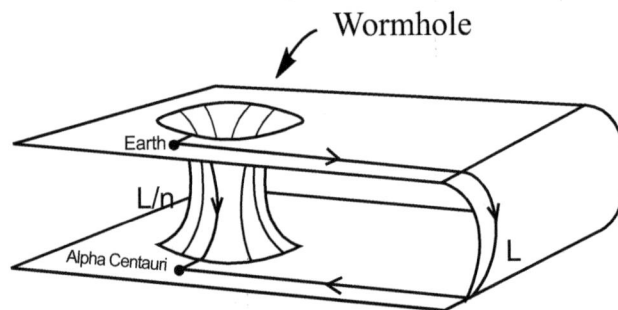

Figure 4: Wormhole creates a shortcut from the Earth to Alpha Centauri

3) <Change time> $L_{star} = c \times (nt)$.

The time "t" in an imaginary time hole is equivalent time of n-fold time in actual space, as shown in Fig. **5**. For example, one second in an imaginary time hole corresponds to one million seconds in actual space.

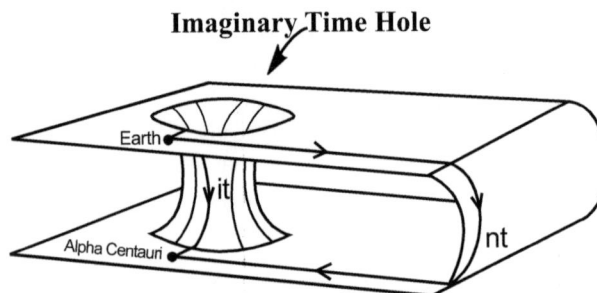

Figure 5: Imaginary time hole creates a shortcut from the Earth to Alpha Centauri

Subsequently, interstellar travel through the imaginary time hole is described as by the following.

Properties of Flat Space

In general, the property of space is characterized by a metric tensor that defines the distance between two points. Here space is divided into two types. Actual physical space that we live in is a Minkowski space, and the world is limited by Special Relativity. It is defined as "Real-Space". Here as a hypothesis, an invariant distance for the time component of Minkowski metric reversal is demanded. This is not a mere time reversal. It is defined as "Hyper-Space". The invariance is identical with the symmetries. Symmetries in nature play many important roles in physics. From this hypothesis, the following arises: the properties of the imaginary time ($x^0=ict$; $i^2=-1$) are required as a necessary result in Hyper-Space. Here, "i" denotes the imaginary unit and "c" denotes the speed of light. The time "t" in Real-Space is changed to imaginary time "it" in Hyper-Space. However, the components of space coordinates(x,y,z) are the same real numbers as the Real-Space. From the above, it is seen that the real time ($x^0=ct$) in Real-Space corresponds to the imaginary time ($x^0=ict$) in Hyper-Space. That is, the following is obtained (see APPENDIX.F and APPENDIX.G):

Real-Space: t (real number), x,y,z (real number) ;

Hyper-Space: it (imaginary number; $i^2=-1$), x,y,z (real number).

The imaginary time direction is at right angles to real time. This arises from the symmetry principle on the time component of Minkowski metric reversal.

Star flight for Stellar System using Imaginary Time Hole

Next, a comparison is made between interstellar travel by Special Relativity and Hyper-Space Navigation. The condition is the same for both cases of navigation, that is, the distance between the earth and the star is 410 light years (*i.e.* Pleiades star cluster) and the velocity of starship is $0.99999c$.

Special Relativity allows the following (see Fig. **6**):

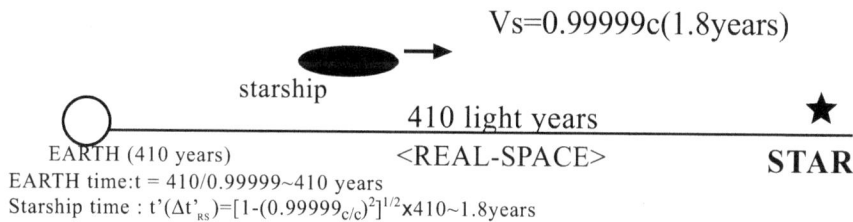

$$Vs=0.99999c(1.8 \text{years})$$

starship

410 light years

EARTH (410 years)   <REAL-SPACE>   **STAR**

EARTH time:t = 410/0.99999~410 years

Starship time : t'(Δt'$_{RS}$)=[1-(0.99999$_{c/c}$)^2]$^{1/2}$x410~1.8years

Figure 6: Interstellar travel by Special relativity

A starship can travel to stars 410 light years distant from us in 1.8 years. However, there exists a large problem as is well known, *i.e.*, the twin, or time, paradox. If the starship travels at a velocity of $0.99999c$, it will arrive at the Pleiades star cluster 1.8 years later. It will seem to the crews in the starship that only 1.8 years have elapsed. But to the people on earth it will have been 410 years. Namely, since the time gap between starship time and earth time is so large, the crew coming back to the earth will find the earth in a different period. This phenomenon is true in our Real-Space. Interstellar travel by this method is non-realistic,*i.e.*, it would just be a one-way trip to the stars.

Hyper-Space Navigation allows the following (see Fig. **7** and APPENDIX.G):

Starship time: $t'(\triangle t'_{HS})$=1.8 years (see Eq.G6)

EARTH time: t($\triangle$t$_{EHS}$)=(1/[1+(0.99999c/c)^2]$^{1/2}$)×1.8=(1/$\sqrt{2}$)×1.8~1.3years (see Eq.G5)

Range: L=0.99999c×1.3×([1+(0.99999c/c)^2]$^{1/2}$/[1-(0.99999c/c)^2]$^{1/2}$) (see Eqs.G7,G9)=0.99999c×1.3×316~410 light years.

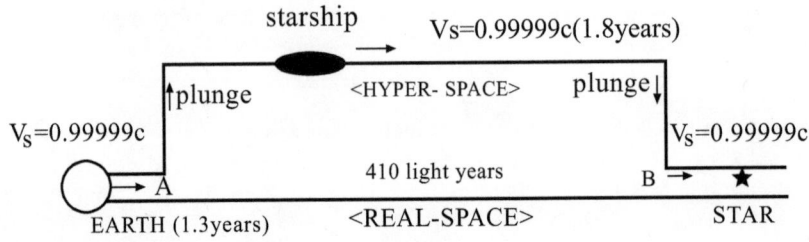

Figure 7: Interstellar travel by Hyper space navigation

A starship can travel to the stars 410 light years distant in 1.8 years. During Hyper-Space navigation of 1.8 years, just 1.3 years have passed on the earth. Therefore, the time gap between starship time and earth time is suppressed. After all, the range and travel time of starship is the same for both kinds of navigation, and travel to the stars 410 light years away can occur in just 1.8 years in both cases. However, by plunging into Hyper-Space featuring an imaginary time, *i.e.* a Euclidean space property, just 1.3 years, not 410 years, have passed on the earth. There is no time gap and no twin or time paradox such as in Special Relativity. Additionally, a starship can travel to the star Sirius 8.7 light years distant us in 0.039 years (14 days). During Hyper-Space navigation of 14 days, just 0.028 years (9 days) have passed on the earth.

Fig. **8** shows such a realistic method for the interstellar travel. In order to reach the target star, the starship which left the Earth at a velocity of approximately 0.1*c* to 0.2*c* moves and escapes completely from the Solar System. After that, the starship is accelerated to nearly the speed of light in Real-Space and plunges into Hyper-Space at point A. In Hyper-Space, the time direction is changed to the imaginary time direction and the imaginary time direction is at right angles to real time. The course of starship is in the same direction, *i.e.* x-axis. With the help of Eqs.(G5), (G7) and (G9) in APPENDIX.G, the crew can calculate the range by the measurement of starship time.

After the calculated time has just elapsed, the starship returns back to Real-Space from Hyper-Space at a point B nearby the stars. Afterward, the starship is decelerated in Real-Space and reaches the target stars. It is immediately seen that the causality principle holds. Indeed, the starship arrives at the destination ahead of ordinary navigation by passing through the tunnel of Hyper-Space. The ratio of tunnel passing time to earth time is 1.4:1 and both times elapse. Hyper-Space navigation can be used at all times and everywhere in Real-Space without any restrictions to the navigation course.

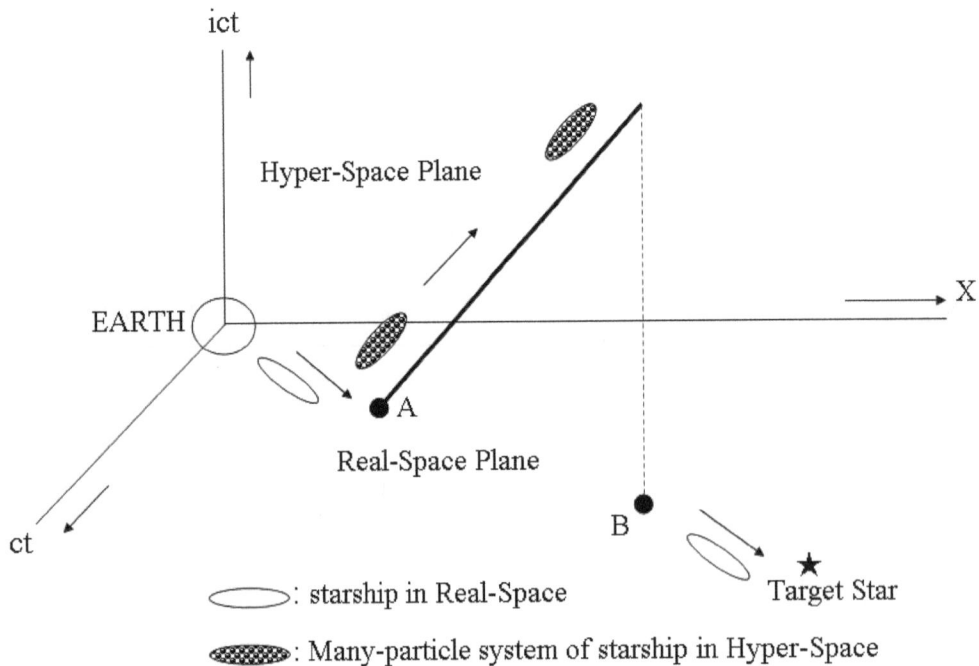

Figure 8: Interstellar travel to the star

This implies that Real-Space always coexists with Hyper-Space as a parallel space. The factor that isolates Real-Space from Hyper-Space consists in the usual-experience "real time" of the former as opposed to the "imaginary time" characterizing the latter. And each space is isolated by the potential barrier (see Fig. G1 in APPENDIX.G).

In general, in case that a diverse two kinds of phase space coexist or adjoin, a potential barrier shall exist to isolate these two kinds of phase space. Starship shall overcome the potential barrier by some methods. One and only difference is either real time or imaginary time. The Real-Space (3 space axes and 1 time axis) and Hyper-Space (3 space axes and 1 imaginary time axis) coexist; the parallel space-time exists as a five dimensional space-time (3space axes and 2 time axes).

Concerning a concept on technical method of plunging into Hyper-Space and returning back to Real-Space, the following study is necessary: 1) Many-Particle Systems for Starship, 2) Wave function of Starship by Path Integrals, 3) Quantum Tunneling Effect, 4) Reduction of Wave function, 5) Starship Information Content Restoring.

While the conceptual framework discussed above is highly speculative, it is in the wake of most of the current international trends on the subject of "Interstellar Travel". Indeed, the problem of interstellar travel consists much more in a navigation theory than in propulsion, as there is no propulsion means, capable of causing a starship to travel at a speed faster than the speed of light.

See APPENDIX.G & H, regarding the concrete plunging method into Hyper-space.

As described above, a plunging into Hyper-Space characterized by imaginary time would make the interstellar travel possible in a short time. We may say that the present theoretical limitation of interstellar travel by Special Relativity is removed. The Hyper-Space navigation theory discussed above would allow a starship to start at any time and from any place for an interstellar travel to the farthest star systems, the whole mission time being within human lifetime.

OTHER POSSIBILITIES FOR FTL INTERSTELLAR TRAVEL

In addition to the proposed theories for faster-than-light (FTL) transportation based on general relativity or string theory, there are other possible methods that would use tachyons or the Casimir effect to create a geometric tunnel equivalent to a wormhole.

The tachyon is a hypothetical particle introduced by Finberg in [4], which was originally invoked to determine the possibility of FTL motion consistent with the Relativity Theory. While the energy and momentum of a tachyon are real numbers, contrary to subluminal particles it has an imaginary mass. Thus its velocity is always greater than the light speed according to special relativity. The mathematics dedicate that a tachyon would behave in a manner opposite to ordinary particles: as a tachyon loses energy it speeds up, and to slow a tachyon requires the addition of energy. If we suppose that there is some sort of quantum tunneling that might allow a subliminal particle to become superluminal (tachyon tunneling), we could then tunnel through such a barrier using quantum mechanics to attain superluminal velocities for space travel.

More generally, T. Musha has shown [14] that highly accelerated elementary particles can transfer into FTL states within a finite time so long as there is no violation of the uncertainty principle of momentum.

There is some experimental evidence indicating the possible existence of FTL phenomena. Measurements on the electron neutrino and the muon neutrino have been coming up with negative values for their rest masses squared [6], which implies that these neutrinos might have imaginary rest masses, thus neutrinos could themselves be examples of tachyons.

Super string theory predicts the existence of FTL modes, shown as follows [17];

For a bosonic string, the world-sheet action is given by:

$$S = \frac{1}{4\pi\alpha'}\int d^2\sigma\sqrt{-g}\,g^{ab}\partial_a X^\mu \partial_b X_\mu, \qquad (4.1)$$

A spectrum of bosonic strings containing a tachyonic mode for open strings becomes

$$m^2 = -\frac{1}{\alpha}, \ 0, \ \frac{1}{\alpha}, \ \frac{2}{\alpha} \ ....., \tag{4.2}$$

which implies the existence of tachyons which have an imaginary mass.

Another possibility was proposed by K. Scharnhorst [16]. According to QED theory, he has shown that the speed of light between the Casimir plates slightly exceeds the speed of light. The energy-depleted vacuum between the Casimir plates is the cause, and increases the speed of light because the electromagnetic wave propagation perpendicular to the plates can be changed by the vacuum structure enforced by the plates. This result suggests that one might conceivably surround a spaceship with a "bubble" of highly energy-depleted (negative energy) vacuum, in which it could travel at FTL speeds, carrying the bubble along with it [3],

Field Propulsion System for Space Travel, 2011, 95-113

CHAPTER 6

Other Exotic Theories and Propulsion Systems

Abstract: There are realms of propulsion theory that stretch over the boundaries of known relativity and quantum theories. Some of these theoretical attempts incorporate electromagnetism or the relativity theory, while others attempt to evolve existing theories. Some of the more interesting theories that may some day be proven to be useful for space travel are presented in this chapter.

INTRODUCTION

There are realms of propulsion theory that stretch the boundaries of known relativity and quantum theories. Some of these theoretical attempts incorporate electromagnetism or the relativity theory, while others attempt to rewrite existing theories. Some of the more interesting theories that may some day be prove to be useful for space travel are presented in this chapter.

GRAVITATIONAL EFFECTS INDUCED BY COUPLING *VIA* SUPERCONDUCTIVITY

In 1992, a Russian material scientist named Dr.Eugene Podkletnov claimed that he had found an antigravity effect while working with a team of researchers at Tampere University of Technology in Finland. They made the device and confirmed a weight loss by using a ring of superconducting ceramics spinning at 5000rpm, which is shown in Fig. **1** [23,27]. The rotating disk showed a variable loss of weight from less than 0.5% to better than 2%.

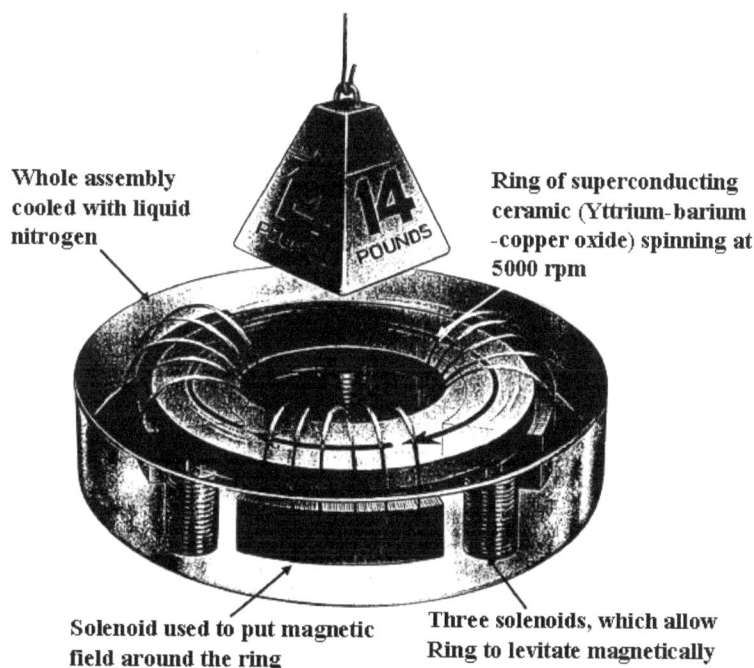

Figure 1: Device to generate gravitational field by using a mass suspended over a superconductor disc

In a follow on to these results, the "Delta-G" experiments were carried by NASA's Marshall Space Flight Center (MSFC) in the USA, which attempted to duplicate Podkletnov's claimed anomalous weight loss in objects of various compositions suspended above a rotating 12-inch diameter Type II ceramic superconductor.

American scientists Douglas Torr and Ning Li, at the University of Alabama, proposed a theory to explain the antigravity effects. [26] This theory can be summarized in a manner parallel to Maxwell's field equations for electromagnetism for the weak field low-velocity limit of general relativity.

If we let E_g be the gravitoelectric field and B_g be the gravitomagnetic field, those then satisfy $E_g = -\nabla \varphi_g - \dfrac{\partial}{\partial t} A_g$

and $B_g = \nabla \times A_g$, where φ_g and A_g are the gravito scalar potential and London gauge potentials.

If we then attempt to apply these equations to the case of rotating superconductors, we can write an equation of motion for a Copper pair of mass m with a charge q moving with the velocity v in electric and magnetic fields E and B, which can be approximated by a generalized Lorentz force as

$$F = q(E + v \times B) + m(E_g + v \times B_g), \tag{1.1}$$

If we then allow that electromagnetism has gravitational counterparts of similar mathematical form, then by analogy Li and Torr obtained:

$$\nabla^2 E = \frac{1}{\lambda_L}\left[E + \frac{m}{q} E_g + \nabla(\varphi + \frac{m}{q}\varphi_g) \right] + \mu_0 \varepsilon \frac{\partial^2 E}{\partial t^2} + \frac{1}{\varepsilon}\nabla \rho_e, \tag{1.2}$$

$$\nabla^2 E_g = -\frac{1}{\lambda_L^2}\frac{\mu_g}{\mu}\frac{m}{q}\left[E + \frac{m}{q} E_g + \nabla(\varphi + \frac{m}{q}\varphi_g) \right] + \mu_{g0}\varepsilon_g \frac{\partial^2 E}{\partial t^2} - \frac{1}{\varepsilon_g}\nabla \rho_m, \tag{1.3}$$

where $\mu_{go} = 1/(\varepsilon_{go} c^2)$ and $\varepsilon_{go} = 1/4\pi G$.

By setting $\tilde{E} = -\partial A / \partial t$ and $\tilde{E}_g = -\partial A_g / \partial t$, they further obtained

$$\tilde{E} + \frac{m}{q}\tilde{E}_g = \left(\frac{\partial A_0}{\partial t} + \frac{m}{q}\frac{\partial A_{g0}}{\partial t} \right) \cdot e^{-z/\lambda}, \tag{1.4}$$

where $\lambda^2 = \lambda_L^2[1 - (\mu_g / \mu)m^2 / q^2]^{-1}$.

In the superconductor satisfying $z \gg \lambda$, this can be written as

$$\tilde{E} + \frac{m}{q}\tilde{E}_g = 0, \tag{1.5}$$

which means that the internal volume of a superconductor is a region of force-free field.

When $\partial A_{g0} / \partial t = 0$, there will be an electrically induced gravitoelectric field given by

$$\tilde{E}_g = \frac{\mu_g}{\mu}\frac{m}{q}\tilde{E}_0, \tag{1.6}$$

When the external magnetic vector potential is ramped up, a time dependent gravitomagnetic vector potential is generated as

$$\tilde{E}_g = -\frac{\partial A_g}{\partial t} = \frac{\mu_g}{\mu}\frac{m}{q}\frac{\partial A_0}{\partial t}, \tag{1.7}$$

This is the prediction of a weight change. Ning Li had predicted one percent weight change in her original calculations.

It can be shown from this that the coherent alignment of lattice ion spins in a rotating type II superconductor will generate a detectable gravitomagnetic effect, and in the presence of a time-dependent applied magnetic vector potential field, a detectable gravitoelectric field.

Experiments conducted for a variety of different test masses near a rotating type II superconductor were tested by applying an electromagnetic field and a weight loss of up to 5% was reported, which shows that there may exist a coupling between rotating superconductors and gravity [16,17]. If this result can but more fully understood theoretically, it could lead to the development of a gravitational propulsion system.

IVANOV'S THEORY ON ELECTROGRAVITICS

Boyko V. Ivanov of the Institute for Nuclear Research and Nuclear Energy in Burgaria obtained a formula for describing coupling between electromagnetism and gravitation from the Weyl-Majumdar-Papapetrou solutions for the metric space-time[10,11,12].

In general relativity, EM fields do indeed alter the metric of space-time and induce a gravitational force through their energy-moment tensor given by

$$T_\nu^\mu = \frac{1}{4\pi}\left(F^{\mu\alpha}F_{\nu\alpha} - \frac{1}{4}\delta_\nu^\mu F^{\alpha\beta}F_{\alpha\beta}\right),$$

(2.1)

where $F_{\mu\nu} = \partial_\mu A_\nu - \partial_\nu A_\mu$ is the electromagnetic tensor and A_μ is the four potential.

We have taken into account that $T_\mu^\mu = 0$. Let us further assume that the metric and E-M fields do not depend on time. In this stationary case, we simplify the problem by setting $A_\mu = (\overline{\varphi},0,0,0)$, then there is just an electric field given by

$$E_\mu = F_{0\mu} = -\overline{\varphi}_\mu,$$

(2.2)

Let us further assume that the space-time is static, so that $f \equiv g_{00} = F(\varphi)$ has the unique form $f = 1 + B\varphi + \varphi^2$, which was found by Weyl in 1917 in the axially-symmetric case, solutions of which are known as Weyl fields.

Then the equation for the gravitational field induced by static electric field can be given by

$$g = c^2 f^{-1}\left(\frac{B'}{2}\sqrt{\frac{\kappa\varepsilon}{8\pi}}\overline{\varphi}_i + \frac{\kappa\varepsilon}{8\pi}\overline{\varphi}\overline{\varphi}_i\right),$$

(2.3)

where $f \equiv g_{00}$, B' is a constant and $\kappa = 8\pi G/c^4$.

From this Ivanov derived the formula of gravitational force F_g shown here for a capacitor as

$$F_g = \sqrt{G\varepsilon}\frac{M}{d}\overline{\psi}_2 = \sqrt{G\varepsilon}\mu S\overline{\psi}_2,$$

(2.4)

where M is the mass of the dielectric, μ is its mass density, ε is dielectric constant, d is the distance between the plates, ψ_2 is the potential of the second plate when $\psi_1 = 0$ and S is an area of the plate. This had been discovered independently by Thomas Townsend Brown (1905-1985) in 1923, together with Prof. P.A. Biefeld in phenomenology that is now referred to as the Biefeld-Brown effect)[13].

This is equivalent (in a modulo Z) to the formula of the force generated by high potential electric field given by Musha shown in Fig. **2**, (derivation of the equation is given in Appendix.I,) again, for a capacitor, which yields

$$F \approx 8.62 \times 10^{-11} Z \mu_0 S \sqrt{\varepsilon_r} V / t \, ,$$

(2.5)

where μ_0 is a total mass of the dielectric per unit area, ε_r is the specific inductive capacity of the dielectric, V is the impressed voltage to the capacitor and t is its thickness.

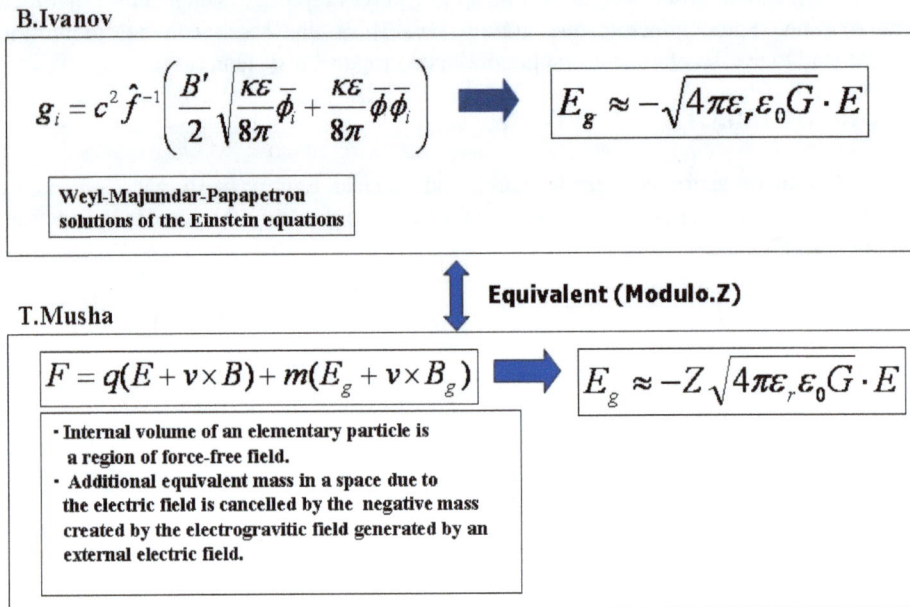

B.Ivanov

$$g_i = c^2 \hat{f}^{-1} \left(\frac{B'}{2} \sqrt{\frac{\kappa \varepsilon}{8\pi}} \overline{\phi}_i + \frac{\kappa \varepsilon}{8\pi} \overline{\phi} \overline{\phi}_i \right) \implies E_g \approx -\sqrt{4\pi \varepsilon_r \varepsilon_0 G} \cdot E$$

Weyl-Majumdar-Papapetrou
solutions of the Einstein equations

Equivalent (Modulo.Z)

T.Musha

$$F = q(E + v \times B) + m(E_g + v \times B_g) \implies E_g \approx -Z \sqrt{4\pi \varepsilon_r \varepsilon_0 G} \cdot E$$

· Internal volume of an elementary particle is
  a region of force-free field.
· Additional equivalent mass in a space due to
  the electric field is cancelled by the negative mass
  created by the electrogravitic field generated by an
  external electric field.

Figure 2: Correspondence between Ivanov theory and the Musha's formula

Ivanov also pointed out that magnetic fields would produce the same effects as electric ones. From the definition of the magnetic field given by

$$H_i = -\frac{1}{2} \sqrt{-g} \varepsilon_{ikl} F^{kl} \, ,$$

(2.6)

Analogous to those for electric fields, Ivanov obtained the formula given by

$$g_i = -\sqrt{G\mu} H_i \, ,$$

(2.7)

where μ denotes the magnetic constant.

Thus according to Ivanov's formulation it is also possible to create an unbalanced acceleration by creating magnetic fields in a ferromagnetic medium. These predicted electromagnetic-gravitational coupling effects would be static and thus they should be able to produce a net force to propel a spaceship.

This effect may be enhanced by combining with other effects, for example, a mass shift due to the interaction of electromagnetic field and the ZPF field, as discussed in Chapter 4. As the magnetic field can be given by $H = \nabla \times A / \mu$, where A is a vector potential of magnetic field, this magneto-gravitational effect can be applied for the propulsion of the spacecraft instead of electrogravitic effect, because the mass shift induced by vector potential enhances the magneto-gravitational effect.

ELECTROMAGNETIC FIELD PROPULSION THEORIES

In establishing the electromagnetic nature of light, Maxwell opposed Weber's "action at a distance" with his dynamical model of a vacuum with hidden matter in motion. His ideas were expanded by Pointing through the energy-flux theorem.

Minkowski found, as a purely mathematical consequence of Maxwell's equations, that the Lorentz force density could be expressed as the divergence of Maxwell's tensor in vacuo, decreased by the rate of change of Poynting's vector as[5]

$$\rho E + \mu_0 J \times H = \nabla \cdot T_{vac} - \varepsilon_0 \mu_0 \frac{\partial}{\partial t} E \times H \,, \tag{3.1}$$

which can be interpreted as a local reaction force acting on charges and currents when the vacuum surrounding them is loaded with electromagnetic momentum.

G.M.Graham and D.G.Lahoz conducted measurement of forces related to electromagnetic momentum by using the cylindrical capacitor with a vacuum gap inside[14,15]. They observed changes in angular momentum that agreed with the classical theory within the error of 20%. This implies that the vacuum can be thought of as possessing its own momentum wherever static fields are set up with non-vanishing Poynting vector. This implies that permanent magnets can be used in conjunction with local electric current flows to build a flywheel that taps into the vacuum electromagnetic energy, for instance that which steadily flows in circles in the vacuum gap of a capacitor.

The notion of EM propulsion by the electromagnetic momentum transfer dates back to J.Slepian, who proposed the EM drive based on $E \times H$ fields. He proposed a concept for electromagnetic spaceship that utilizes the vacuum momentum under electromagnetic field, shown as[1,24]

$$F = \frac{\partial D}{\partial t} \times B - \varepsilon_0 \mu_0 \frac{\partial E}{\partial t} \times H \tag{3.2}$$

or

$$F = \frac{\partial D}{\partial t} \times B - \varepsilon_0 \frac{\partial E}{\partial t} \times B \,. \tag{3.3}$$

Fig. **3** shows the apparatus composed of a parallel plate capacitor between two solenoids electrically wired in series, which was suggested as a driving mechanism for a spaceship. He supposed that this device, composed of inductive-capacitive system supplied with $10^{8.}$ Hz power, would produce a thrust of magnitude 1.5×10^7 dynes for the case $H = 5000$ gauss, $I = 1000$ Amperes, and a plate separation of $0.03m$, if properly phased [25].

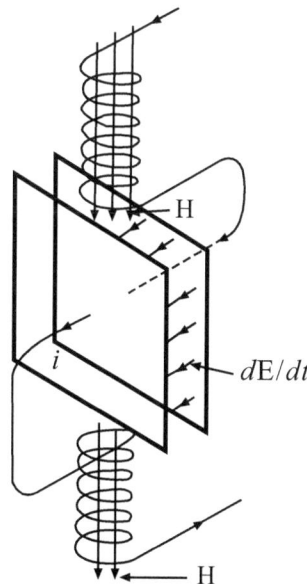

Figure 3: Schematic diagram of the Slepian's drive

Prof. Mario Pinheiro at Institute Superior Tecnico in Lisbon has proposed a new approach for EM propulsion *via* vacuum electrodynamics based on a fluidic viewpoint[18,19]. His approach establishes an analogy or isomorphism between the systems of equations governing electromagnetic field and those governing fluid dynamics. Table **1** shows the correspondence of field variables in electromagnetism and hydrodynamics [18]. From this analogy the force acting on a moving charge with the velocity v becomes

$$F = -\rho \frac{\partial}{\partial t} A - \nabla[\rho(v \cdot A) + \rho\varphi] + \rho[v \times B] \,, \qquad (3.4)$$

which is similar to Euler and Bernoulli's integrals of the fluid dynamic equations. This can be applied for a plasma fluid with the charge density ρ. In particular, Eq.6.3.4 predicts the existence of longitudinal forces in the EM field, a prediction that has yet to be experimentally explored. This same equation predicts the feasibility of the electromagnetic propulsion device by utilizing an unbalanced force arising from an electromagnetic field in the vacuum.

Table.1: Correspondence of field variables in electromagnetism and hydrodynamics[19]

Electromagnetism	Hydrodynamics
$\rho(r,t)/\varepsilon_0$	Hydrodynamic charge $n(r,t)$
Permeability of the vacuum μ_0	Mass density ρ
Electric potential $\varphi(r,t)$	Massic enthalpy $p/\rho(r,t)$
Scalar potential χ	Velocity potential Φ
Vector potential $\mathbf{A}(r,t)$	Velocity (or hydrodynamic momentum) $\mathbf{u}(r,t)$
Electric field $\mathbf{E}(r,t)$	Lamb vector $\mathbf{l}(r,t)$
Magnetic field $\mathbf{B}(r,t)$	Vorticity $\boldsymbol{\omega}(r,t)$
Voltage (or electric tension) $U(r,t)$	$p(r,t) = \rho(r,t)\Phi(r,t) + \rho(r,t)u(r,t)^2/2$
Electric current $\mathbf{I}$	Circulation Γ
Electromotive force	Hydrodynamic force
$(\mathbf{E} = -\dfrac{\partial \mathbf{A}}{\partial t} - \nabla\varphi - \nabla(\mathbf{v}\cdot\mathbf{A}) + [\mathbf{v}\times\mathbf{B}])$	$(\mathbf{E}_H = -\dfrac{\partial \mathbf{u}}{\partial t} - \nabla\left(\dfrac{p}{\rho} + \dfrac{u^2}{2}\right) + [\mathbf{u}\times\boldsymbol{\omega}])$

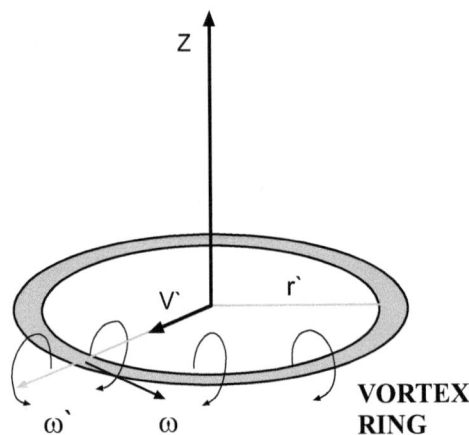

Figure 4: Concept of Electromagnetoroid

Based on this framework, Pinheiro [22] introduced the concept of electromagnetoroid, showing in Fig. **4**. Such as structure could be thought of as being powered by Abraham's force. Abraham's force is generally driven by some external source, e.g. the gas fall to star center in polar jets, or a device forcing a plasma in a spiral trajectory to a central axle. Abraham's force represents a kind of vortex structure formed in the physical vacuum and the related counteraction propels any material structure through space-time. By utilizing this effect, Pinheiro presented a space propulsion method in which a strong electric current or charged plasma is fed to the center of a spiral as shown in Fig. **5**.

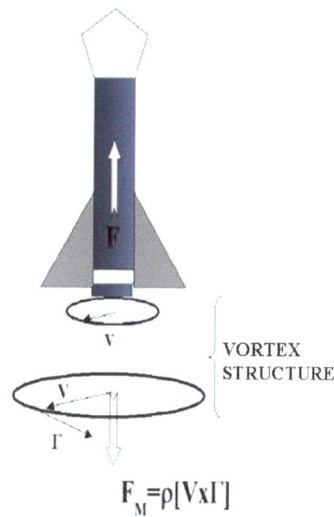

$$F_M = \rho[V \times \Gamma]$$

Figure 5: Space propulsion method by Electromagnetoroid

At the core of the vortex structure the resultant force is aligned to the axis given by $F = \rho\left[\vec{v} \times \vec{\Gamma}\right]$. It can also be shown that Abraham's force is the analog of the Magnus force.

HEIM'S THEORY ON THE GRAVITOPHOTON PROPULSION

The Heim's quantum theory of gravity is based on the geometric view of Einstein, which is that altered geometry itself is the cause of all force interactions. However Heim used the structure of Einstein's field equations only as a template for describing force interactions in a higher dimensional discrete space, and extended them to the quantum level. Heim's theory requires extending spacetime with extra dimensions, an eight dimensional discrete space in which a smallest elemental surfaces exists[3]. Within the quantum spacetime of Heim's theory, elementary particles are represented as "hermetry forms" or multidimensional structures of space. Heim claimed that his theory yields particle masses directly from fundamental physical constants and that the resulting masses are in agreement with experiment. For Heim, this composite nature was an expression of internal, six-dimensional structure. After his death, others have continued with his multi-dimensional "quantum hyperspace" framework[4].

Although it purports to unify quantum mechanics and gravitation, the original Heim's theory cannot be considered a theory of everything because it does not incorporate all known experimental data. In particular, it gives predictions only for properties of individual particles, without making detailed predictions about how they interact. The theory also allows for particle states that don't exist in the Standard Model, including a neutral electron and two extra light neutrinos, and many other extra states. Presently, there is no known mechanism for the exclusion of these extra particles, nor an explanation for their non-observation.

Predictions claimed to have been derived from first principles that are experimentally testable are: 1) predictions of the masses of neutrinos; 2) predictions of new particles; 3) predictions of excited states of existing particles, and 4) predictions for the conversion of photons into the so-called "gravito-photons" resulting in a measurable force. Heim's theory predicts that it should be possible to produce an effective "gravity" force by converting photons into "gravito-photons," a particle which he predicts the existence of but has not yet been directly observed.

In the 1950s, Heim had predicted what he termed a 'contrabary' effect whereby photons, under the influence of a strong magnetic field in a certain configuration, could be transformed into 'gravito-photons', which would provide an artificial gravity force.

As of late 2006, groups at Berkeley and elsewhere were attempting to reproduce this effect. By applying their 'gravito-photon' theory to bosons, Dröscher and Häuser were able to predict the size and direction of the effect.

Heim attempted to resolve incompatibilities between quantum theory and general relativity. To meet that goal, he developed a mathematical approach based on quantizing space-time itself, and proposed the "metron" as a (two-dimensional) quantum of (multidimensional) space.

The Heim's theory is further developed by Walter Dröscher and Jochem Häuser.

According to their paper, the fifth interaction, termed as graviton force, would accelerate a material body without the need of propellant. Gravitophoton interaction is a gravitation-like force that come into both attractive and repulsive. Gravitophoton particles are produced in pairs from vacuum itself by the effect of vacuum polarization (virtual electrons), under the presence of a very strong magnetic field.

For gravitation, Heim theory assumes that a gravitational potential arises from the gradient of a field $\varphi(r)$. Position dependent mass is the function m, and r is the radial distance from a quanta of a point mass.

A differential equation used to describe the basis is

$$\left(\frac{d\varphi}{dr}\right)^2 + 32\frac{c^2}{3}F\left(\frac{d\varphi}{dr} + F\varphi\right) = 0 \,, \tag{4.1}$$

$$F = \frac{1}{r}\frac{h^2 + \gamma\ m^3 r}{h^2 - \gamma\ m^3 r} \tag{4.2}$$

If this equation is nondimensionalized the characteristic length of the system is

$$r_c = \frac{h^2}{\gamma\ m^3} \,, \tag{4.3}$$

The characteristic length is the distance from a point mass for which the field $\varphi(r) = 0$. It is also where the field attains its absolute minimum. Hence, the gravitational force is identically zero at this distance.

The solution to the differential equation has the curve $\varphi(r)$ concave up. The gravitational potential that arises from this field can be positive, negative or zero.

From this the Heim-Lorentz equation can be obtained as

$$F_{gp} = -\Lambda_p e\mu_0 v^T \times H \,, \tag{4.4}$$

which describes the gravitational interaction resulting from negative gravitophotons, where the index p in Λ_p indicates that only photon and neutron absorption processes are considered, v^T is the velocity of a rotating torus of current, and H is the magnetic field generated by the current loop.

Gravitophoton propulsion takes place in two phases. In phase one, a spaceship is subject to acceleration in 4-dimensional space-time. Acceleration is achieved by the absorption of negative gravitophotons through the photons and neutrons in the torus material. In order to cover large interplanetary distances a transition into parallel space would be required, which is phase two of the field propulsion, involving the repulsive quintessence particle. A transition into a parallel space leads to an increase in speed by a factor n, compared to our 4-dimentional space-time.

The value n is obtained by

$$n = (g_{gp}^+ / g_g) \cdot (G_{gp} / G) \,, \tag{4.5}$$

where g_{gp}^+ is the field strength of the gravitophoton field, g_g is the gravitational field strength and G_{gp} is a coupling constant of gravitophotons.

In the rotating torus of Fig. **6**, the positive and negative gravitophoton fields are generated together, and, because of energy conservation, their strength are equal and can be directly calculated from Eq.(4.4)

The Heim-Lorenz force equation can produce a force component in the vertical direction to the rotating torus as shown in Fig. **6** [24].

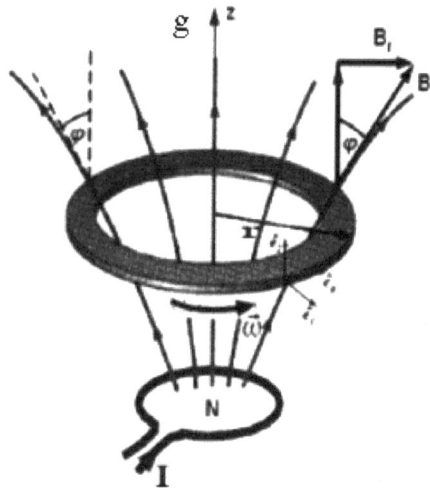

Figure 6: Physical principle of the device to generate a gravitational field

If his theoretical prediction is correct, a workable space propulsion device with enough lift itself directly from the surface of the Earth could be realized.

For instance, it has been estimated that an acceleration becomes larger than 1g for a spacecraft with the mass of 1.5×10^5 kg, a magnetic induction of 20T, a rotational speed of the torus of $v^T = 10^3$ m/s, and a torus mass of 2×10^3 kg, which could result in a total flight time to the Moon on the order of 4 hours [4].

HAYASAKA'S THEORY ON THE DE RHAM COHOMOLOGICAL EFFECT OF FOUR DIMENSIONAL ANGULAR MOMENTUM

It was found that the clockwise spinning of a gyro (made from non-magnetic materials) around the vertical axis causes weight decrease [6] and fall-time increase [8]. The gravitational acceleration decrease on the clockwise spinning is proportional to the product of a coefficient θ (=7×10^{-14}m^{-1}), velocity of light c, and rotational velocity rω, where r is rotational radius, and ω is angular frequency. This is due to the generation of torsion field on the de Rham cohomology effect of four-dimensional angular momentum of a rotational object.

The gravitational moments with time on both rotations of a mass point in opposite directions along a loop (on the base) are in an equivalent class with de Rham cohomology group relating to the invariant angular momentum on a mirror transformation. The de Rham cohomology effect causes the torsion fields of different strength in both rotations, and then there is the possibility of the parity breaking of gravitational force and the generation of topologically repulsive forces in an object's spinning.

Gravitational repulsive force is generated due to the de Rham cohomology effect of four-dimensional angular momentum; by means of the clockwise circulation flow of magnetic fluid where the vast positive energy of magnons consisting of spin waves occurs. Repulsive force generator is constructed by two toroidal tubes in which ferromagnetic fluids applied by external magnetic fields and laser beams are circulated in the reverse directions each other [9].

GRAVITATIONAL REPULSIVE FORCE DUE TO THE DE RHAM COHOMOLOGICAL EFFECT ON 4-DIMENSIONAL ANGULAR MOMENTUM OF AN OBJECT'S CLOCKWISE SPINNING

It was anticipated that the gravitational force caused by an object's clockwise rotation in viewing from the above is different from that of the counter-clockwise rotation from the de Rham cohomology second theorem, and that the clockwise rotation causes gravitational repulsive force due to torsion field [7].

According to the de Rham cohomology theorem [2], if the integrals of two quantities Ω and Ω' are equal along a closed path C, there is a difference provided by an exact differential $d\chi$ between Ω and Ω', that is,

$$if \oint_C (\Omega - \Omega') = 0, \quad then \quad \Omega - \Omega' = d\chi \neq 0. \tag{5.1}$$

To discuss the above-mentioned theorem for the concrete application, let us consider a rotor which is constrained on a horizontal plane, and is spinning in a stationary state around the vertical axis. The rotor is regarded as only a dust ensemble consisting of many mass points, so that the problem of the stationary spinning motions of the rotor is reduced to the problem of the stationary rotations of a mass point along a circle. The calculations of the gravitational forces associated with both rotational motions are carried out in an approximately flat space-time where the Cartesian coordinates are set by x^0, x^1, x^2 and x^3 (ct, x, y and z). In the case of a closed system, the 4-dimentional angular momentum $M^{\mu\nu}$ is given by

$$M^{\mu\nu} = \frac{1}{c} \int (x^\mu f^\nu - x^\nu f^\mu) dx^0. \tag{5.2}$$

If the theorem is applied to the conserved 4-dimensional angular momentum M^{03} or M^{30} of a rotating object, there is certain difference between the gravitational forces on the clockwise and the counterclockwise rotations, as follows. We are now concerned with x^3 component of gravitational force f^μ or f^ν. Therefore, only the component of angular momentum M^{03} or M^{30} of a rotating object given by the following will be considered;

$$M^{03}(L) = \frac{1}{c} \int \{x^0 f^3(L) - x^3 f^0(L)\} dx^0 = \frac{1}{c} \int \{x^0 f^3(R) - x^3 f^0(R)\} dx^0 = M^{03}(R). \tag{5.3}$$

From Eq.(5.3), taking the invariance of M^{03} or M^{30} on the mirror transformation into account, we get

$$M^{03}(L) - M^{03}(R) = 0, \quad then \quad \frac{1}{c} \oint_C [x^0 \{f^3(L) - f^3(R)\} - x^3 \{f^0(L) - f^0(R)\}] dx^0 = 0, \tag{5.4}$$

where the C on integral denotes a closed path in the base space.

Applying Eq.(5.1) to Eq.(5.4), we get

$$\frac{1}{c}[x^0 \{f^3(L) - f^3(R)\} - x^3 \{f^0(L) - f^0(R)\}] dx^0 = d\chi \neq 0. \tag{5.5}$$

Therefore, the conservation of $M^{03}(L)$ on the counterclockwise rotation and $M^{03}(R)$ on the clockwise rotation leads to the following representation

$$\frac{1}{c} \oint_C \{x^0 f^3(L) - x^3 f^0(L)\} dx^0 - \frac{1}{c} \oint_C \{x^0 f^3(R) - x^3 f^0(R)\} dx^0 = \oint_C d\chi. \tag{5.6}$$

Since arbitrary function χ is given generally by Fourier expansion, $d\chi$ is satisfied by the following periodic function

$$d\chi = \{-\sum_N A_N N \overset{*}{\omega} \sin N \overset{*}{\omega} x^0 + \sum_N B_N N \overset{*}{\omega} \cos N \overset{*}{\omega} x^0\} dx^0 \ . \tag{5.7}$$

From Eqs.(5.5), (5.6) and (5.7), we obtain

$$f^3(L) - f^3(R) = -\frac{c\sum_N A_N N \overset{*}{\omega} \sin N \overset{*}{\omega} x^0}{x^0} \ . \tag{5.8}$$

The second term on Eq.(5.7), *i.e.* $B_N N \overset{*}{\omega} \cos N \overset{*}{\omega} x^0 / x^0$ is not be accepted because of its divergence to infinity for x_0 →0. Applying the de Rham cohomology theorem to 4-dimensional angular momentum, it is no longer $f^3(L) = f^3(R)$, and there exists a finite difference. Since each term of the right hand side on Eq.(5.8) represents a sampling function or Dirac's δ function for $N>>1$, the right hand side is equal to the sum of pulse functions with respect to time x^0. This means that an object rotating along a closed path C causes the excitation of the vacuum in atom. In the other words, both rotations cause the different gravitational fields due to the topological effect. It means that the connection coefficient on both rotations is not symmetrical ($\Gamma^\mu_{\nu\sigma} \neq \Gamma^\mu_{\sigma\nu}$), and then the fields are torsion like or twisted. From the analogy with β-decay (*i.e.*, weak interaction) in which the clockwise circulating electrons in a coil generating external magnetic field and the emitted electrons from nucleus form the left-handedness, the gravitational repulsive acceleration α(R) due to the topological effect which is caused by only the clockwise rotation of a gyro constructed by non-magnetic materials is given by

$$\alpha(R) = \theta c r \omega = 7 \times 10^{-14} \times c r \omega = 2 \times 10^{-5} r \omega \ (m/s^2) \ . \tag{5.9}$$

Equation (5.9) has been confirmed by both experiments of weight change and fall-time measurements.

As to a propulsive engine generating an excited space, an engine system which supplies topological effect induced by means of the clockwise circulation flow of magnetic fluid is considered. This engine system yields the topologically excited space. The principle of engine is based on de Rham cohomology effect of four-dimensional angular momentum by means of the clockwise flow of magnetic fluid where occurs the excitations of magnons consisting of spin waves. The required value of magnetic field is 0.1 Tesla to 2-3 Tesla..

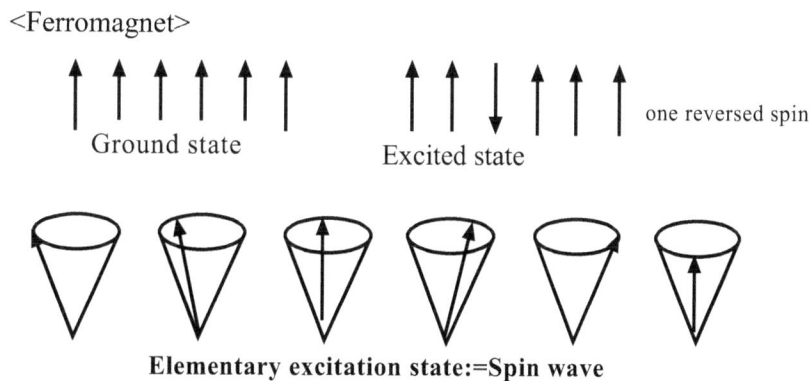

Figure 7: Magnons spin wave

In general, concentration of energy is required to excite the space as vacuum. The excitation of space demands some "CATALYSIS". The catalysis assists the interaction between Fermion field (matter field) and Boson field (vacuum field). Magnons (quasi-boson particles) as spin waves are considered to be effective as to catalysis. Magnon is a quantized spin wave. In the ground state of a simple ferromagnet, all spins are parallel. As a possible excitation, one spin is reversed. The low-lying elementary excitations are spin waves. The ends of the spin vectors precess on the surfaces of cones, with successive spins advanced in phase by a constant angle. Thus a reversed spin generates a spin wave, *i.e.*, a magnon. Magnon is a quasi-particle and obeys Bose-Einstein statistics (Fig. 7).

REPULSIVE FORCE GENERATOR DUE TO TOPOLOGICAL EFFECT OF CIRCULATING MAGNETIC FLUID

Repulsive force generator is constructed by two toroidal tubes in which ferromagnetic fluids applied by external magnetic fields and laser beams are circulated in the reverse directions each other. The generator yields the topologically excited space around the spaceship.

The lifting acceleration acting on all the atoms constructing the generator due to the topological effect, *i.e.* α is given to be

$$\alpha = \theta c R \omega \approx 6.4 \times 10^{-7} \times \frac{sn}{2\pi R} c R \omega = 1.9 \times 10^{2} \, snv \quad (m/s^{2}) . \tag{5.10}$$

Here, v is the number of circulations of magnetic fluid per second. The acceleration α does not depend on the radius of toroidal tube. Although the other mass constructing the spaceship except the mass of generator is large, the acceleration of topologically gravitational repulsive force is very large as understood on the clockwise spinning gyro constructed by non-magnetic materials. Therefore, spaceship can be lifted easily from celestial body.

In fact, α can be made easily larger than 9.8 m/s^2 of the Earth through the quantity snv. The α is the gravitational repulsive acceleration induced by the generator, which is the engine system due to the topological effect of circulating magnetic fluid. The estimation of gravitational repulsive acceleration is made for the magnons generated in the domains applied by external magnetic fields. If the magnons generated in the processes of parallel pumpings of laser beams are taken into account, the repulsive acceleration is larger.

NAVIGATION METHOD FOR A DESTINATION STAR

We have several navigation methods for a destination star. The lifting force directs along the vertical axis on the celestial body. Here, the standard axis of a spinning gyro is settled along the vertical direction on the celestial body to be starting point. For the case of which the spaceship arrives at a space point of zero potential field (*i.e.* flat space), the general navigation method is as follows (Fig. **8**). The circular plain formed by the toroidal tubes is inclined toward the destination star, and the magnon generations in the forward domains are actualized in strong magnitude, while the backward domains are operated in weak magnitude. Since there occurs potential difference between the forward and the backward parts, the propulsive force is generated toward the destination star. The magnitude of propulsive force can be regulated by way of magnitude change of magnon energy through the changes of the number of domains n, and the number of circulations v on Eq.(5.10), and also through the change of intensity of external magnetic fields.

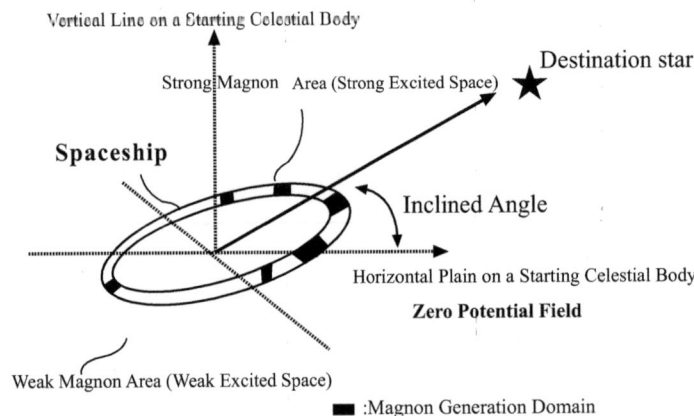

Figure 8: Arrangement of Magnon generation for the general navigation toward a destination star

For the case where the azimuth direction of spaceship in the circular plain formed by the toroidal tubes is changed, this case is realized by the shifts of magnon generation domains along the circumferences of the toroidal tubes. For the case where the spaceship is arriving near the surface of the destination star, magnon generation is decreased gradually in the balance of the positive potential generated by spaceship and the negative potential of the destination star. For all the case mentioned above, the gravitational positive field generated by generator installed in spaceship must be independent of that of the surrounding space. From this, the generator is operated in prompt on-off control.

Superstring Based Field Propulsion System

It has been a long-standing challenge for theoretical physics to construct a theory of quantum gravity. In recent years, concern has been raised about superstring theory. Superstring theory is the leading candidate for a quantum theory of gravity. By the way, there are five kinds of superstring theories, *i.e.* Type I, Type IIA, Type IIB, SO(32) heterotic, $E_8 \times E_8$ heterotic. The five different self-consistent superstring theories are nothing but different solutions of a single theory, called "M-theory". In this revised picture, the various string theories are nothing but different vacua of a single theory. M-theory indicates that the "true home" of the theory may actually be the eleventh dimension, where we find new, exotic objects, such as super membranes.

Recently there has been remarkable progress in the understanding of some string solitons called "D-branes". Superstring theory contains "D-brane" solitons that are extended membranes of various possible spatial dimensions. Now, the nonperturbative regime of superstring theory yields entirely new types of membranes, called "D-branes", which play an essential role in the theory. Open strings with Dirichlet boundary conditions can end on these D-branes.

Superstring theory is not a theory of strings after all; indeed, there are extended objects in the theory ('D-branes') which carry the basic charges of a special class of higher rank antisymmetric fields.

In 2004, Minami proposed a Superstring-Based Field Propulsion Concept [20].

The starting conjecture is that superstrings might – as a whole – behave like the polymer chains of some elastic bodies (e.g. the rubber). It is well known that, for such type of elasticity, the related force is given by the product of temperature and entropy gradient. It is explained how a pulsed engine, capable to excite space and induce entropy gradient, might move a spaceship. The entropy can be calculated from the number of microstates of the polymer chains at a given temperature. From superstring theory, the Schwarzschild radius of string increases with the string coupling constant; therefore, strings might be black holes. The second key-point consists in applying the 5-dimensional Reissner-Nordström black hole solution to a configuration of strings and solitons. By using the quantum counting method, one can calculate the entropy of a D-brane in terms of the number of states. The third basic assumption is that space-time has entropy through D-branes, which are supposed to cling it. This suggests that the structure of space-time is also composed of some kinds of physical microstates and offers the properties of entropy.

Since the D-brane is a deformable and dynamical object, and couples with gravity, it should be a source to generate the curvature of space-time. Therefore, the theory of D-brane is applicable to the field propulsion principle utilizing the properties of the field of space-time. The propulsion system induced by entropy gradient of field based on D-brane is introduced. It seems that the entropy is dependent on the fabric of space-time. Since the entropy which conforms to the entropy of statistical mechanics is induced by D-brane, the physical substance has started to become clear. This propulsion principle is linked to the fabric of space-time.

The strings of superstring theory are considered as the threads of the space-time fabric. In a sense, it is as if individual strings are the "shards" of space-time, and only when they appropriately undergo sympathetic vibrations, the conventional notions of space-time emerge. Accordingly, string seems to be fundamental element of the substructures of space-time. This indicates that strings might behave like the polymer chains of some elastic bodies like rubber.

Gravitational Field Equation Induced by Superstring Theory

Since a massless spin-2 particle (*i.e.*, graviton) occurs in the spectrum of the closed string, superstring theory essentially includes a gravitational field. Consequently, Einstein's gravitational field equation is obtained from not only superstring theory but also D-brane solution of superstring theory.

Here, the solution based on D-brane is stated. Considering the gravitational theory unified with arbitrary antisymmetric tensor field and dilaton field, the general string action for gravity coupled to a dilaton φ and m different n_A-form field strengths is

$$S = \frac{1}{16\pi G_D} \int d^D x \sqrt{-g} \left[R - \frac{1}{2}(\partial\varphi)^2 - \sum_{A=1}^{m} \frac{1}{2 \cdot n_A!} e^{aA\varphi} F_{nA}^2 \right],$$
(6.1)

where G_D is the D dimensional Newton constant, n_A is the degree of forms F, φ is the dilaton field, a_A is the constant, and F_{nA} is the strength of field.

This action describes the bosonic part of D=11 or D=10 supergravties.

In superstring theory, dilaton φ, which is scalar field, determines the strength of the string self-interaction, $g_S = e^\phi$. The dilaton field is akin to the gravitational field.

The field equation derived from the above-stated action is obtained by varying with respect to the metric tensor $g^{\mu\nu}$ as follows:

$$R_{\mu\nu} - \frac{1}{2} g_{\mu\nu} R = \frac{1}{2} [\partial_\mu \varphi \partial_\nu \varphi - \frac{1}{2} g_{\mu\nu} (\partial\varphi)^2] + \sum_{A=1}^{m} \frac{1}{2 \cdot n_A!} e^{aA\varphi} [n_A (F_{nA})^2{}_{\mu\nu} - \frac{1}{2} g_{\mu\nu} F_{nA}^2].$$
(6.2)

Eq.(6.2) gives the gravitational field equation (6.3) regarding the right side of Eq.(6.2) as matter field.

$$R^{\mu\nu} - \frac{1}{2} g^{\mu\nu} R = -\frac{8\pi G}{c^4} T^{\mu\nu},$$
(6.3)

where $R^{\mu\nu}$ is the Ricci tensor, R is the scalar curvature, G is the Newton's constant, c is the speed of light, and $T^{\mu\nu}$ is the energy momentum tensor.

The right side of Eq.(6.2) is the energy momentum tensor $T^{\mu\nu}$. This indicates that Eq.(6.2) includes the D-brane solution. Thus, the gravitational field equation is derived from D-brane in superstring theory at low energy effective action.

Propulsion System Utilizing the Elastic force Induced by Entropy Gradient

In general, elasticity has two kinds of nature, that is, energy elasticity (crystalline elasticity) like spring and entropy elasticity (rubber elasticity) like rubber. Energy elasticity is due to the deformation of interatomic distance or displacement between molecules. It corresponds to the decrease of internal energy. Entropy elasticity (rubber elasticity) is due to thermal motion of the polymer chains. It corresponds to an increase of entropy. Elasticity of rubber is very different from that of crystallic solids. The elastic constant of rubber increases with temperature.

The space as vacuum is considered to preserve the properties of entropy elasticity from the viewpoint of the latest cosmology and theory of elementary particles.

Further, the space as vacuum is considered that the entropy of space-time can be defined as an assembling of strings and strings as the constituents of space-time correspond to the polymer chains in the elastic body.

Fig. **9** shows the propulsion principle utilizing the entropy gradient. When the engine of spaceship is switched on, the space in a surrounding region of spaceship is excited with increased entropy. The entropy of space in the rear vicinity of spaceship is the same as the usual entropy of space; therefore an entropy gradient is produced in the direction of engine as shown in Fig. **9**.

This state is equivalent to the state of stretched rubber, and is the state in which the stretched rubber (*i.e.*, small entropy) returns or contracts to the original natural state through the restoring force.

S$_{large}$: Entropy is large

S$_{small}$: Entropy is small

S$_{small}$

● : Engine

F

S$_{small}$ F   S$_{large}$ Spaceship

S$_{large}$

F   S$_{small}$

F   Excited Space

S$_{small}$

$$F = T \frac{\partial S}{\partial r}$$ (T: Temerature density, S: Entropy)

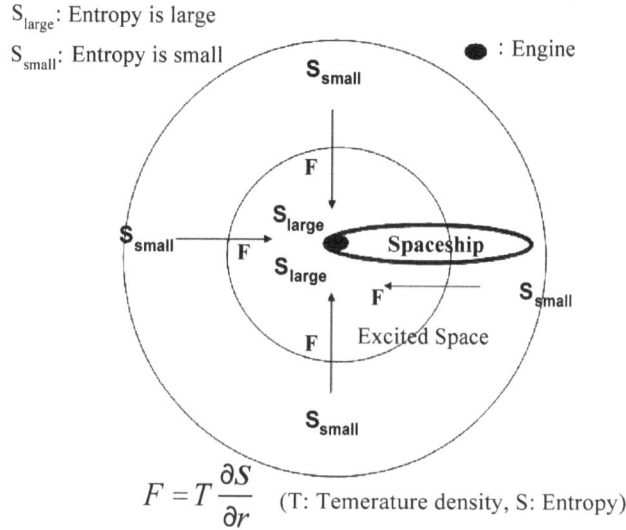

Figure 9: Propulsion principle induced by Entropy gradient

As shown in the usual rubber elasticity, the elastic force is induced by entropy gradient in the direction of increasing entropy (from small entropy to large entropy). The spaceship is propelled receiving the elastic force "*F*" from the field of entropy gradient.

The excited space shows the property of rubber elasticity, and as is well known in the statistical theory of rubber elasticity, the elastic force F is given by

$$F = T \frac{\partial S}{\partial r},$$ (6.4)

where T is the temperature density as energy density, S is the entropy, r is the distance from engine.

From the statistical mechanics, entropy S is obtained as follows:

$$S = k \log W ,$$ (6.5)

where W is the number of microstates, k is the Boltzmann's constant.

Substituting Eq.(6.5) into Eq.(6.4), we get

$$F = kT \frac{\partial \log W}{\partial r} .$$ (6.6)

The field of elastic force, that is, acceleration field is generated around the spaceship. In the on-state of engine (denoted as ● in Fig. **9**), the large entropy state is generated in the front vicinity of spaceship. In the rear vicinity of spaceship apart from engine, the entropy is small as usual natural space. Accordingly, there exists an entropy gradient in the range of excited space. In this state, the elastic force is generated in the direction of increasing entropy (from small entropy to large entropy), and this state corresponds to the small entropy of stretched rubber. However, spaceship cannot move as long as the engine generates the entropy gradient in the surrounding area of spaceship, due to the interaction between entropy gradient field of space and spaceship. This is because an action of engine on space (*i.e.*, generating entropy gradient field) is in equilibrium with the elastic force as reaction from space. In the state of engine on, it generates the excited space, *i.e.* entropy gradient around the engine of spaceship. Although the excited space as entropy gradient contracts toward the engine, the excited space cannot contract. This state is maintained till the engine off, due to the equilibrium state.

It is consequently necessary to shut off the equilibrium state in order to actually move the spaceship. As a continuum, the space has a finite strain rate, *i.e.* speed of light. When the engine stops generating the entropy gradient field in space, a finite interval of time is necessary for the generated entropy gradient field to return to ordinary space. In the meantime, the spaceship is independent of entropy gradient field in space. It is therefore possible for the spaceship to proceed ahead receiving the elastic force from the space. In the state of engine off, the spaceship is dragged as the excited space (*i.e.*, entropy gradient field) contracts, so the spaceship can be propelled forward. This state corresponds to the stretched rubber (*i.e.*, small entropy) contracting to initial state (*i.e.*, large entropy) of rubber. As mentioned above, since the engine must necessarily be shut off for propulsion, the spaceship can get continuous thrust by repeating the alternate ON/OFF change in the engine operation at high frequency.

Black Hole Entropy in D-brane

For the above-mentioned propulsion principle induced by entropy gradient to be valid, the structure of space as a vacuum must possess the fabric of corresponding to arbitrary physical state quantities offering the entropy. In this section, the number of microstates "W" is derived from the fine structure of space using the concept regarding Black Hole Entropy in D-brane.

Now as is well known, the quantum mechanics implies that black holes emit thermal radiation with a temperature obeying the first law of thermodynamics, as

$$dM = T_H dS,$$ (6.7)

and the black holes have an entropy equal to one quarter of their horizon area in Planck units, as

$$S = \frac{A_H}{4G_N} \quad (= \frac{A_H}{4G_N \hbar} kc^3),$$ (6.8)

where M is the black hole mass, T_H is the black hole temperature, S is the entropy, A_H is the horizon area, G_N is the Newton's constant, k is the Boltzmann's constant, and c is the speed of light (from now on $\hbar = k = c = 1$).

Black holes have entropies proportional to the area of horizon. However, the microscopic degrees of freedom that give rise to the entropy are not visible in the classical theory. Recently, the number of states turns out to be precisely the exponential of the Bekenstein-Hawking entropy. The description uses properties of some superstring theory solitons called D-branes.

In most physical systems, the thermodynamic entropy has a statistical interpretation in terms of counting microscopic configurations with the same macroscopic properties, and in most cases this counting requires an understanding of the quantum degrees of freedom of the system.

In superstring theory, the fundamental object is a string, and the string oscillates. The eigen frequencies of strings are all considered as kinds of particles. All particles are excitations of one dimensional extended object - a string. There is fundamental length scale l_s called the string scale which is set by the string tension $T = 1/2\pi \cdot l_s^2 = 1/2\pi\alpha'$. Newton's constant is related to l_s by $G_N \approx g_s^2 l_s^2$ (in four dimensions) where g_s is the string coupling constant.

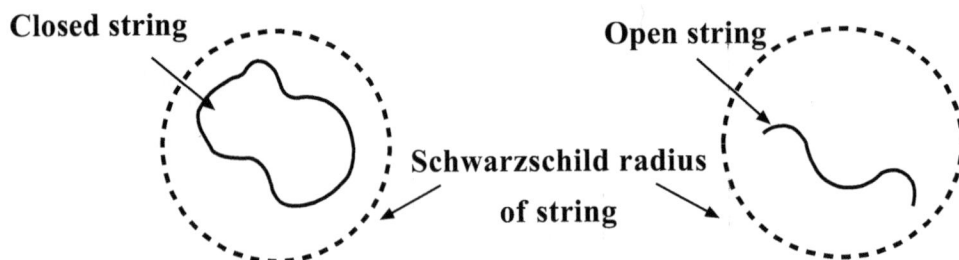

Figure 10: Black hole of String

The gravitational field produced by these states depends on G_N and increases with the string coupling. Accordingly, Schwarzschild radius of string $r_s = 2GM/c^2$ becomes larger than string size itself by increasing with the string coupling constant. Hence, the string becomes black hole. In general, in the case that Schwarzschild radius of massive body becomes larger than its radius; the massive body becomes black hole. The string becomes the black hole by gravitational collapse as shown in Fig. **10**.

Since the strings having same massive states are degenerate, all strings become a black hole. From the no-hair theorem of black hole, the black hole which has arbitrary mass is just one. Accordingly, the string state having the same mass becomes the same black hole. Since the entropy is given by the logarithm of degeneracy of system, the black hole entropy is considered as the logarithm of degeneracy of eigen frequency of strings. The black hole entropy induced by statistical mechanics is based on above-stated fundamental notion. Namely, to make the string into black hole by adjusting the string coupling constant yields that the black hole entropy can be solved as string entropy; the number of string states agrees with the black hole entropy. This entropy precisely counts the number of string states with the given energy and charges. String is the candidate for black hole. Charged black holes in General Relativity are characterized by their mass M and charge Q. The black hole with $Q=M$ is called extrenmal, since it has the minimum possible mass for a given charge.

The solution with the minimum mass for a given charge ($Q=M$) will then be BPS. The so called BPS solutions preserve some supersymmetries. As the string coupling is increased, the gravitational field becomes stronger and the metric becomes an extreme black hole - now with finite horizon area. Each BPS state with the same Q at weak coupling is described by the same black hole at strong coupling. The BPS state, that is, the relation of $Q=M$ is invariant for the string coupling constant; the degeneracy of systems is also invariant. The entropy can be accurately evaluated regardless of strong coupling problem by using BPS state.

There also exists the BPS state in D-brane. D-branes are dynamical objects and the open string states describe their fluctuations. The oscillations of branes are described by massless open strings with both ends attached to the same brane. There are also a large number of massless states coming from open strings with ends attached to different branes. Especially, an excited brane corresponds to having a gas of these open strings on the brane. The numbers of states of D-branes can be counted at weak coupling state like as strings. The mass of D-brane is given by $M \approx g_s^2 l_s^2$, and then the mass of Q D-branes is given by $M \approx Q/g_s l_s$. Using $G \approx g_s^2 l_s^2$, the Schwarzschild radius of D-brane ($r_D = 2GM/c^2$) becomes the order of $r_D \approx g_s l_s Q$ (setting c=1). Accordingly, the large string coupling constant (g_s) yields that D-brane also becomes black hole.

Now first, the entropy in classical black hole solutions shall be obtained. The charged black holes in five space-time dimensions are given by the Reissner-Nordström solution. The five-dimensional Reissner-Nordström black hole is a solution of the five-dimensional Einstein plus Maxwell action. There is a configuration of strings and solitons in the type II B string theory on $M^5 \times T^5$ ($T^5 = T^4 \times S^1$) that can be identified with the Reissner-Nordström black hole [21].

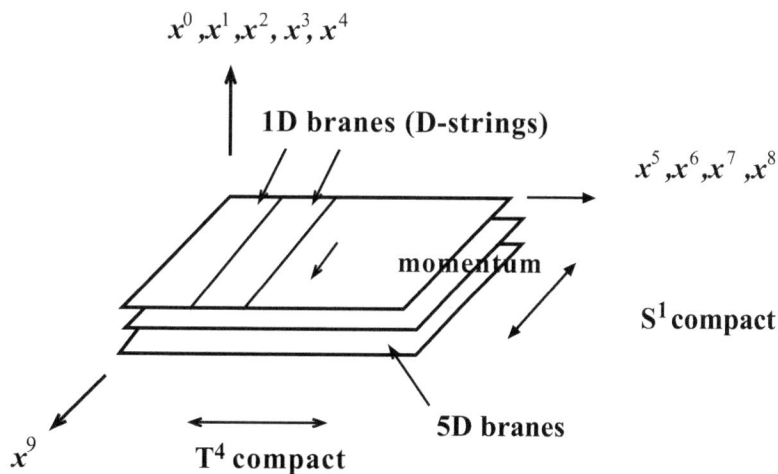

Figure 11: Configuration of D-brane

We denote the M^5 coordinates by $(x^0, x^1, x^2, x^3, x^4)$ and the torus coordinates by $(x^5, x^6, x^7, x^8, x^9)$ as shown in Fig. **11**.

Type II B compactified on T^5 (four-dimensional torus $T^4 \times$ circle S^1) is considered. The configuration is constructed as follows: first wrap a number Q_5 of 5D-branes on the torus (T^5), then wrap Q_1 1D-branes (D-strings) along one of the directions of the torus (in the direction x^9). In addition, some momentum $P_9=n/R_9$ (R_9 is the radius of the circle) along the 1D-branes, *i.e.* in the direction x^9. The 1D-branes are embedded in 5D-branes. Q_1 is a Ramond-Ramond electric charge, and counts the 1D-branes. Q_5 is a magnetic charge, and counts the number of 5D-branes. The third charge, n, corresponds to the total momentum ($P_9=n/R_9$) along the branes in the direction x^9. It turns out that a 5D-branes which wraps once around the five torus carries charge $Q_5 =1$, and a D-string (1D-branes) wrapped once around the circle with radius R carries charge $Q_1=1$. The charges Q_1, Q_5 and n are naturally integer quantized.

As to the type II B string theory on T^5 ($T^5 = T^4 \times S^1$), a configuration of Q_5 D-five branes wrapping the whole T^5, Q_1 D-strings wrapping the S^1 and momentum n/R_9 along S^1, choosing S^1 to be in the direction x^9 are considered. All charges n, Q_1, Q_5 are integers.

The entropy of five dimensional extremal Reissner-Nordström black hole is obtained as follows (see APPENDIX.K):

$$S_{BH} = \frac{A}{4G_5} = 2\pi\sqrt{Q_1 Q_5 n} \ . \tag{6.9}$$

What has been argued above is semi-classical, the statistical mechanical origin of entropy being ignored fully. Namely, it consists in recasting known results in terms of the string background equations.

It is now required to find a statistical derivation of this relationship, *i.e.* Eq.(6.9), using quantum counting method. The key observation is that how to count the BPS D-brane states is already known.

As mentioned previously, the interactions of D-branes is given by open string which stretch between them. The counting of states is then dominated by open strings which are free to move in the other four dimensions of the four-torus. Fortunately, the counting of string states is well known.

The asymptotic formula for the degeneracy of level n states in a conformal field theory of central charge c is given by

$$d_n \approx \exp\left(2\pi\sqrt{\frac{nc}{6}}\right) . \tag{6.10}$$

In a simple conformal field theory with species of bosons and fermions, central charge c is given by $c=4Q_1 Q_5$ $(1+1/2) =6Q_1 Q_5$. Thus, since the statistical entropy is the logarithm of the degeneracy d_n ($S=log\ d_n$), using Eq.(6.10), we get

$$S_{BH} = \log d_n = 2\pi\sqrt{6Q_1 Q_5 n / 6} = 2\pi\sqrt{Q_1 Q_5 n} , \tag{6.11}$$

in perfect agreement with Eq.(6.9), thereby confirming that $S = A_H / 4G_N$. This implies that the number of states of D-brane is related to the entropy and there exists a fine structure in space-time. The thermal random perturbation of fine structure of space-time seems to yield the entropy.

Although above stated black hole solution is in five dimensions, the more realistic case of four dimensional black holes also yields the same result. As described above, the statistical mechanics interpretation of black hole was enabled by using D-brane in superstring theory.

Entropy Structure of Space

In the first section, the propulsion principle induced by entropy gradient is studied. This principle is based on considering space-time as a kind of elastic field like rubber. Consequently, as is well known in the statistical theory of rubber elasticity, the elastic force is generated by the entropy gradient. Since the elastic force is a body force, there is no action of inertial force.

Substituting Eq.(6.11) into Eq.(6.4), the following elastic force is obtained

$$F = 2\pi T \frac{\partial (Q_1 Q_5 n)^{\frac{1}{2}}}{\partial r}.$$

(6.12)

Next, in the second section, the Bekenstein-Hawking entropy is derived for a class of five-dimensional extremal black holes in string theory by counting the degeneracy of BPS soliton bound states. The Bekenstein-Hawking entropy was initially discovered between the laws of black hole dynamics and the laws of thermodynamics. Therefore, the Bekenstein-Hawking entropy is just analogy and behaves in every way like thermodynamic entropy not statistical entropy.

A precise statistical mechanical interpretation of black hole entropy – including the numerical factor – by counting black hole microstates has been required. Supposing that the string or D-brane is the constituent of space-time is suggestive of the existence of possible quantum states for the black hole's space-time. This indicates that entropy of space-time can be defined as an assembling of strings. Strings as the constituents of space-time correspond to the polymer chains in the elastic body. Since the statistical entropy is the logarithm of the number of states (*i.e.*, degeneracy of system), it is necessary to consider what kinds of physical state exist.

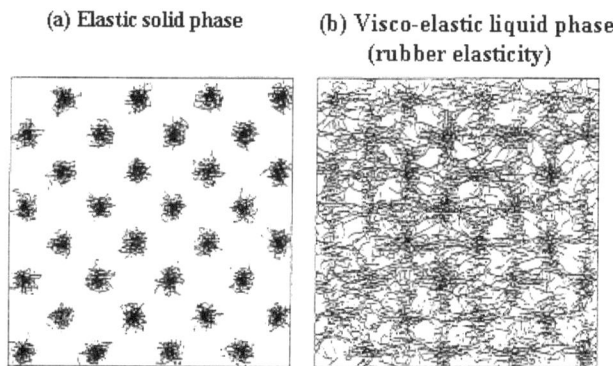

(a) Elastic solid phase   **(b) Visco-elastic liquid phase (rubber elasticity)**

Figure 12: Fine structure of Space-Time

Fig. **12** shows that the open strings cling to the field of space. Fig. **12** (**a**) shows the state of present cosmic space in ultra-low temperature, and Fig. **12** (**b**) shows the state of the early universe in ultra-high temperature.

The excitation of space implies that the ordered phase of open strings clung to space in Fig. **12** (**a**) is transferred to the disordered phase of open strings clung to space in Fig. **12** (**b**) by some trigger. It corresponds to that the number of twined open strings is transferred from ordered phase (small entropy) to disordered phase (large entropy). This picture indicates that these states can be interpreted as entropy.

Accordingly, it suggests the possibility of a new propulsion theory induced by entropy gradient.

Conclusions

A number of researchers in the USA, Japan and Europe are applying advanced concepts and methods of Relativity and Quantum Field Theory to investigate new ways of accomplishing space travel apart from the conventional methods.

Field propulsion system described in this book as a new propulsion method must obey the conservation laws in nature, such as conservation of energy, and conservation of momentum. Since the field propulsion does not expel any momentum (i.e., there is no reaction mass from spaceship), it appears to violate the conservation law of momentum. However, the essential underlying principle of field propulsion is: space is not a state of absolute void or nothingness. Space is a physical entity, or has some physical structure as a continuum which can be re-acted against under the right conditions, that is, the surrounding space has to react, and this structure of space may be a result of any or all of the fundamental forces of nature.

Concerning the conservation of energy, discovery of new energy methods to power these propulsion devices is required.

In general, a spaceship (mass of M) traveling at a speed V needs the kinetic energy of $E_K = \frac{1}{2} MV^2$. For instance, a spaceship traveling at a speed equal to 0.1 c has a specific kinetic energy equal to 450 Telajoules (TJ) per kilogram (of spaceship mass).

The required energy of spaceship of 100 ton at a speed of 0.1 c is 4.5×10^{19} Joules. Any propulsion system, i.e. not only conventional propulsion but also field propulsion, its power source must provide huge energies, that is, $E = Pt$ (P is power in watts, t (s) is acceleration and deceleration time). This energy problem is common to all propulsion systems if the high speed is required.

In the case of interstellar travel, the main problem could be summarized in one word: Time. Therefore, ultra high-speed is required; as a consequence, huge energy is required. Especially, field propulsion systems require huge energy source due to their performance of high acceleration and high speed.

How can we produce such a huge energy?

Since we did not address the energy source for field propulsion in this book, we mention here about "New Energy Sources for Interstellar Travel" in brief.

For example, let consider the ZPF propulsion system;

From the equation given by

$$P_{field} = Z\sqrt{4\pi\varepsilon G}\, \frac{M}{v_d}\sqrt{\frac{P}{\omega C}}\left(1 + \frac{3\pi}{16}\frac{e^2 G N^2 R}{\varepsilon_0^2 mc^6 \omega_e l^2}\frac{P}{\omega C}\right),$$

which is the relation between the generated momentum P_{field} and the electric power P supplied to the propulsion device shown in Chapter. 4.

Supposing that the propulsion device to produce acceleration has the dimension of 5m with the radius and 10m with the length, then the maximum electric power to attain the speed 9.5×10^5 km/h, that is about 1 % of the light speed, becomes $P \approx 20$ Tera watts, that is huge electric power which cannot be obtained by the conventional electric generator. This is almost equal to average power usage world wide for all counties on the planet and thus the powerful energy source must be developed for the ZPF propulsion of this spaceship. This is tremendous electric power, which cannot be generated by conventional electric generators.

Takaaki Musha and Yoshinari Minami

Forward demonstrated in his paper that the Casimir force could in principle be used to extract energy from ZPF field in the vacuum. He showed that any pair of conducting plate at close distance experiences an attractive Casimir force that is due to electromagnetic ZPF field of the vacuum. Thus a vacuum-fluctuation battery can be constructed by utilizing the Casimir effect to do work on a stack of charged conducting plate.

Cole and Puthoff verified that energy extraction from the ZPF field is not contradictory to the laws of thermodynamics. For thermodynamically irreversible processes, heat can be produced and made to flow, either at $T = 0$ or at any other $T > 0$ situation [1, 2]. However, if one is considering a net cyclical process on the basis, the energy would not be able to be continually extracted without a violation of the second law of thermodynamics. Thus, Forwards vacuum-fluctuation battery cannot be cycled to yield a continuous extraction of energy.

Another example of a method for extracting from the ZPF field is a patent by Mead and Nachamkin. They proposed that a set of resonant dielectric spheres be used to extract energy from the ZPF field and convert it into electric power [3].

Although novel energy extraction mechanism have been proposed in some literatures, no practical technique has been demonstrated in the laboratory at the present stage.

Minami proposed "Liquid Metal MHD Power Generation System Using Antiproton Annihilation Reactor" [4, 5].

As a simple trial calculation, the $1 m^3$ of hydrogen in the standard state contain 2.7×10^{25} hydrogen molecules, namely, 5.4×10^{25} protons. Provided that the same number of protons and antiprotons exist in $1 m^3$, the energy of $6.9 \times 10^{15} J / m^3 = 1.9 \times 10^9 kWh / m^3$ ($1kWh = 3.6 \times 10^6 J$) is obtained from the annihilation reaction. The heat source gained from an antiparticle annihilation reactor is supplied to an MHD (Magneto hydrodynamics) generator utilizing electromagnetic induction effect of an electromagnetic fluid.

Antiprotons are being produced today by using accelerators such as bevatron not only in Japan but also in various countries in the word, but the amount of production is extremely little. The new methods of production of antiprotons and storage technology to avoid collision with residual gas and material are required.

Furthermore, the vacuum energy or so-called dark energy may be useful in the near future.

We presented the underlying principles that exhibit the current progress of the space propulsion physics from science journal and conferences. The goal of a human beings' space travel in the 21st century needs to attain to beyond even the stellar system not only a solar system. By developing propulsion physics further, it is required to make the new propulsion technology of get ahead of the limit of the existing propulsion system. The domain where the present physics is not yet completed is left, and it is expected that practical development of an epoch-making new propulsion system and the navigation theory will be made with fast development of these physics.

Light speed travel would make us to explore the Universe far away from the Earth

References

CHAPTER 2

[1]   Kuninaka H. *et al.*, Powered Flight of HAYABUSA in Deep Space, Paper AIAA 2006-4318, 43rd AIAA/ASME/SAE/ASEE Joint Propulsion Conference, 2006.
[2]   Larson C.W., Mead Jr. F.B., Myrabo L.N., Messitt D.G., Paper AIAA 2001-0646, 39th AIAA Aerospace Sciences Meeting & Exhibit. 8-11 January 2001, Reno, NV.
[3]   Mead Jr. F.B. and Myrabo L.N., Flight Experiments and Evolutionary Development of a Laser Propelled, Trans-Atmospheric Vehicle, STAIF-98 Congress, Albuquerque(NM), USA, Jan.25-98, 1998.
[4]   Minami.Y, Uchida.S, Wate Vapor Propulsion Powered by a High-Power Laser-Diode, *JBIS*, 62, 2009: 332-339.
[5]   Myrabo L.N. and Mead,Jr., F.B., Ground and Flight Tests of a Laser Propelled Vehicle, AIAA98-1001, 36th Aerospace Sciences Meeting & Exhibit, January 12-15, 1998/ Reno, NV.
[6]   Niino M., Minami Y., Nakamura T., Study and Research Result of Advanced Space Propulsion Investigation Committee (ASPIC), Paper IAA-97-IAA.4.1.09, presented at the 48th IAF Congress, 1997.
[7]   Phipps C.R., Michaelis M. M., LISP: Laser impulse space propulsion, Laser and Particle Beams, 12 (1), 1994: 23-54.
[8]   Tsujikawa Y., Imasaki K., Niino M., Minami.Y, Hatsuda Y., Japanese Activity on the Laser Application in Space , presented at the AHPLA' 99, Osaka University, Japan, 1999.

CHAPTER 3

[1]   Alcubierre M., The Warp Drive: Hyper-Fast Travel within General Relativity, *Class. Quantum Gravity*, 11, L73-L77, 1994.
[2]   Forward R.L., Etracting electrical energy from the vacuum by cohesion of charged foliated conductors, *Physical Review B*, 30 (4), 1984: 1700-1702.
[3]   Froning Jr. H.D., Vacuum Energy For Power and Propulsive Flight?, AIAA 94-3348, 30th AIAA/ASME/SAE/ASEE Joint Propulsion Conference, June 27-29, 1994 / Indianapolis, IN, USA.
[4]   Fronings Jr H.D., Barrett T.W., Inerrtia Reduction-And Possibly Implusion-By Conditioning Electromagnetic Fields, AIAA 97-3170,33rd AIAA/ASME/SAE/ASEE, Joint Propulsion Conference & Exhibit, July 6-9, 1997.
[5]   Kaku M., Introduction to Superstrings and M-Theory 2nd ed, Springer, 1999.
[6]   Kane G., Modern Elementary Particle Physics, Addison-Wesley, 1993.
[7]   McMahon D., Quantum Field Theory Demystified, McGraw-Hill , New York, 2008.
[8]   Michael E.P., Daniel V.S., An introduction to Quantum Field Theory, Perseus Books, 1995.
[9]   Minami Y., Space Strain Propulsion System, 16th International Symposium on Space Technology and Science (16th ISTS), Vol.1, 1988: 125-136.
[10]   Minami Y., Possibility of Space Drive Propulsion, paper IAA-94-IAA.4.1.658, presented at 45th IAF Congress, 1994.
[11]   Minami,Y., An Introduction to Concepts of Field Propulsion, *JBIS*, 56, 2003: 350-359.
[12]   Millis M.G., EXPLORING THE NOTION OF SPACE COUPLING PROPULSION, In Vision 21: Space Travel for the Next Millennium, Symposium Proceedings, Apr 1990, NASA-CP-10059, 1990: 307-316.
[13]   Millis M.G., Williamson G.S., NASA Breakthrough Propulsion Physics Workshop Proceedings, NASA/CP-1999-208694, January 1999: 263-273.
[14]   Milonni P.W., Cook R.J., Goggin M.E., Radiation pressure from the vacuum: Physical interpretation of the Casimir force, *Phys. Rev. A*, 38 (3), 1988: 1621-1623.
[15]   Niino M., Minami Y., Nakamura T., Study and Research Result of Advanced Space Propulsion Investigation Committee (ASPIC), paper IAA-97-IAA.4.1.09, presented at the 48th IAF Congress, 1997.
[16]   Polchinski J., String Theoruy, Vol.1,2, Cambridge University Press, Cambridge,1998.
[17]   Puthoff H.E., Little S.R., Ibison M., Engineering the Zero-Point Field and Polarizablel Vacuum for Interstellar Flight, *JBIS*, 55, 2002: 137-144.
[18]   Yam P., Exploiting Zero-Point Energy, *Scientific American,* Dec. 1997: 82-85.

CHAPTER 4

[1]   Alcubierre M., The warp drive: hyper-fast travel within general relativity, *Class. Quantum Grav.*,11:1994, L73-L77.

[2]   Braginsky V.B. *et al.*, Laboratory experiments to test relativistic gravity, *Physical Review D*, 15 (8), 1977: 2047-2068.

[3]   Boxman, R. L., Martin, P. J., Sanders, D. M., Handbook of Vacuum Arc Science and Technology: Fundamentals and Applications, Noyes Publications,1996.

[4]   Caves, C.M., Quantum-mechanical noise in an interferometer, *Phys.Rev.D.*, 23, 1981: 1693-1708.

[5]   Davis, E. W., Puthoff, H. E., Experimental Concepts for Generating Negative Energy in the Laboratory, in proceedings of Space Technology and Applications International Forum (STAIF-2006), edited by M. S. El-Genk, AIP Conference Proceedings, Melville, New York, 2006: 1362-1373.

[6]   Drummond I.T., Hathrell S.J., QED vacuum polarization in the background gravitational field and its effect on the velocity of photons, *Phys.Rev.D.* 22, 1980: 343-355.

[7]   Feynman, R. P., Leighton, R. B., Sands, M., The Feynman Lectures on Physics, Vol. II, Addison-Wesley, New York, 1964.

[8]   Flügge W., Tensor Analysis and Continuum Mechanics, Springer-Verlag Berlin Heidelberg New York, 1972.

[9]   Forward R.L., Antigravity, Proceedings of the IRE, 961: 1442.

[10]   Forward R.L., Guidelines to Antigravity, *American Journal of Physics*, 37, 1963: 166-170.

[11]   Forward R.L.(Forward Unlimited, Malibu CA), Letter to Minami, Y. (NEC Space Development Div., Yokohama JAPAN) about Minami's "Concept of Space Strain Propulsion System", (17 March 1988).

[12]   Froning Jr H.D., Vacuum Energy for Power and Propulsive Flight?, 30th AIAA/ASME/SAE/ASEE Joint Propulsion Conference, AIAA, 1994: 1-15.

[13]   Froning Jr. H.D. and Barrett, T.W., Space Coupling by Specially Conditioned Electromagnetic Fields, in proceedings of Space Technology and Applications International Forum (STAIF-98), 1998: 1449-1454.

[14]   Fung Y.C., Classical and Computational Solid Mechanics, World Scientific Publishing Co. Pre. Ltd., 2001.

[15]   Gueron E., Adventures in Curved Spacetime, *Scientific American*, August 2009: 26-33.

[16]   Harris E.G., Analogy between general relativity and electromagnetism for slowly moving particle in weak gravitational fields, *American Journal of Physics*, 59 (5), 1991.

[17]   Haisch B., Rueda, R., Puthoff, H. E., Inertia as a zero-point-field Lorenz force, *Phys. Rev. A*, 49 (2), 1994: 678-694.

[18]   Haisch, B., Rueda, A., The Zero-Point Field and the NASA Challenge to Create the Space Drive, NASA Breakthrough Propulsion Physics Workshop, NASA Lewis Res. Ctr., Aug.12-14: 1997.

[19]   Jefimenko O.D., Causality, Electromagnetic Induction and Gravitation, Electric Scientific Company, Star City, USA,1992.

[20]   Kuo C-I., Ford, L. H., Semiclassical gravity theory and quantum fluctuations, *Phys. Rev. D*. 47, 1993: 4510-4519.

[21]   Long K.F, The Status of The Warp Drive, *JBIS*, 61, 2008: 247-352.

[22]   Longo M.J., Swimming in Newtonian space-time: Orbital changes by cyclic changes in body shape, *American Journal of Physics*, 72 (10), 2004: 1312-1315.

[23]   Mahood T.L. "Propellantless Propulsion: Recent Experimental Results Exploiting Transient Mass Modification," in Proceedings of Space Technology and Applications International Forum (STAIF-99), edited by M.S. El-Genk, AIP Conference Proceedings 458, Melville, NY, 1999: 1014-1020.

[24]   Matloff G. L., Deep Space Probes, Springer, 2000; page 127 (Ch. 9: 9.4 'CABBAGES AND KINGS': GENERAL RELATIVITY AND SPACETIME WARPS.

[25]   Minami,Y., Possibility of Space Drive Propulsion (IAA-94-IAA.4.1.658), 45th IAF Congress, 1994.

[26]   Minami Y., Space Drive Force Induced by a Controlled Cosmological Constant, paper IAA-96-IAA.4.1.08, presented at 47th IAF Congress, 1996.

[27]   Minami Y., Spacefaring to The Farthest Shores - Theory and Technology of A Space Drive Propulsion System, *JBIS*, 50, 1997: 263-276.

[28]   Minami Y., Space Strain Propulsion System, 16th International Symposium on Space Technology and Science (16th ISTS), Vol.1, 1988: 125-136.

[29]   Minami Y., Conceptual Design of Space Drive Propulsion System, STAIF-98, edited by Mohamed S. El-Genk, AIP Conference Proceedings 420, Part Three, 1516-1526, Jan.25-29, 1998, Albuquerque, NM, USA.

[30]   Minami Y., Extraction of Thrust from Quantum Vacuum Using Squeezed Light, STAIF-2007, edited by Mohamed S. El-Genk, AIP Conference Proceedings, Feb.11-15, 2007, Albuquerque, NM, USA.

[31]   Minami,Y., Preliminary Theoretical Considerations For Getting Thrust Via Squeezed Vacuum, *JBIS*, 61, 2008: 315-321.

[32]   Millis M.G., Exploring The Notion Of Space Coupling Ppropulsion, In Vision 21: Space Travel for the Next Millennium, Symposium Proceedings, Apr 1990, NASA-CP-10059, 1990: 307-316.

[33]   Millis M.G., Challenge to Create the Space Drive, *Journal of Propulsion and Power*, 13 (5), 1997: 577-582.

[34]   Musha T., Theoretical Explanation of the Biefeld-Brown Effect, *Electric Spacecraft Journal*, Issue.31, 2000:29, also in Iwanaga N., Review of Some Field Propulsion Methods from the General Relativistic Standpoint, Space Technology and Applications International Forum-1999, 1999: 1051-1059.

[35] Musha T., Explanation of dynamical Biefeld-Brown Effect from the standpoint of ZPF field, Proceedings of International Academy of Astronautics, IAA, Aosta, Italy, 2007; and also in Musha,T., Explanation of dynamical Biefeld-Brown Effect from the standpoint of ZPF field, JBIS, 61(9), 2008: 379-384.

[36] Musha T., Possibility of the Space Propulsion System Utilizing the ZPF Field, Space, Propulsion & Energy Sciences International Forum-SPESIF-2009, AIP, 2009: 194-201.

[37] Puthoff, H. E., Gravity as a zero-point-fluctuation force, *Phys. Rev. A*, 39 (5), 1989: 2333-2342.

[38] Puthoff, H. E., Can the vacuum be engineered for spaceflight applications? Overview of theory and experiments, *Journal of Scientific Exploration*, 12 (1), 1998: 295-302.

[39] Van Den Broeck C., Alcubierre's Warp Drive: Problems and Prospects, STAIF-2000, edited by Mohamed S. El-Genk, CP504, pp.1105-1110, Jan.31-Feb.4,2000, Albuquerque, NM, USA.

[40] Weigert S., Spatial squeezing of the vacuum and the Casimir effect, *Phys. Lett A.*, 214,1996: 215-220.

[41] Wisdom J., Swimming in Spacetime: Motion by Cyclic Changes in Body Shape, *Science*, 299, 2003: 1865-1869.

[42] Zampino E.J., Critical Problems for Interstellar Propulsion Systems, Available from : ralph.open − aerospace.org/deep/repository/zampino2.pdf ; website shown on Google, June 1998.

CHAPTER 5

[1] Alcubierre M., The warp drive: hyper-fast travel within general relativity, *Class. Quantum Grav.*, 11:1994, L73-L77.

[2] Benett G.L, Forward R.L., Frisbee R.H., Report on the NASA/JPL Workshop on Advanced Quantum/Relativity Theory Propulsion, 31st AIAA/ASME/SAE/ASEE Jouint Propulsion Conference and Exhibit, AIAA, 1995: 1-20.

[3] Cramer J.G., NASA Goes FTL Part 1: Wormhole Physics, *Analog Science Fiction and Fact*, Vol.CXIV, No.15, 1994: 118-122.

[4] Feinberg G., Possibility of Faster-Than-Light Particles, *Physical Review*, 159(5), 1967: 1089-1100.

[5] Forward R.L., Space Warps: A Review of One Form of Propulsionless Transport, *JBIS*, 42, 1989: 533-542.

[6] Forward R.L., Faster-Than-Light, *Analog Science Fiction and Fact*, Vol.CVX, No.4, 1995: 30-50.

[7] Froning Jr. H.D., Requirement For Rapid Transport To The Further Stars" *JBIS*, 36, 1983: 227-230.

[8] Hawking S., A Brief History of Time, Bantam Publishing Company, New York, 1988.

[9] Hawking S., Hawking on Bigbang and Black Holes, World Scientific, Singapore, 1993.

[10] Minami,Y., Hyper-Space Navigation Hypothesis for Interstellar Exploration (IAA.4.1-93-712), 44th Congress of the International Astronautical Federation(IAF),1993.

[11] Minami,Y., Travelling to the Stars: Possibilities Given by a Spacetime Featuring Imaginary Time, *JBIS*, .56, 2003: 205-211.

[12] Minami,Y., A Perspective of Practical Interstellar Exploration: Using Field Propulsion and Hyper-Space Navigation Theory in the proceedings of Space Technology and Applications International Forum (STAIF-2005), edited by M. S. El-Genk, AIP Conference Proceedings 746, Melville, New York, 2005; 1419-1429.

[13] Morris, S.M., Thorne, K.S., Wormholes in spacetime and their use for interstellar travel: A tool for teaching general relativity, *American Journal of Physics*, 56, 1988: 395-412.

[14] Musha T., Possible existence of faster-than-light phenomena for highly accelerated elementary particles, *Speculations in Science and Technology*, 21, 1998:29-36.

[15] Santoli S., Nano-to-Micro Integrated Single-Electron Bio-Macro Molecular Electronics for Miniaturised Robotic "Untethered Flying Observers", *Acta Astronautica*, 41, Nos. 4-10, 1997: 279-287.

[16] Scharnhorst K., On propagation of light in the vacuum between plates, *Physics Letters B*, 236(3), 1990: 354-359.

[17] Szabo R.J, An Introduction to String Theory and D-brane Dynamics, Imerial College Press, London, 2004.

[18] Visser M., Lorenzian Wormholes: From Einstein to Hawking, American Institute of Physics, New York, 1996.

[19] Vulpetti G., Problems and Perspectives in Interstellar Exploration, *JBIS*, 52, 1999: 307-323.

CHAPTER 6

[1] Corum J.F., Dering, J.P, Pesavento P., Donne A., EM Stress-Tensor Space Drive, Proceedings of Space Technology and Applications International Forum-1999, American Institute of Physics, 1999: 1027-1032.

[2] De Rham, G. "Variétés Différentiables", Hermann, Paris,1960.

[3] Droscher W., Hauser J., Guidline for a Space propulsion devise based on Heim's Quantum Theory, AIAA 2004-3700, AIAA/ASME/SAE/ASE, Joint Propulsion Conference & Exibit., Frorida, July, 2004: 28-56.

[4] Droscher W., Hauser J., Heim Quantum Theory for Space Propulsion Physics, Space Technology and Applications International

From-STAIF, 2005:1430-1440., also available from: www.hpcc.space/ publications/documents/heim_staif2005-letter.pdf.

[5] Graham G.M., Lahoz D.G., Observation of static electromagnetic angular momentum in vacuo, *Nature*, 285 (15), 1980: 154-155.

[6] Hayasaka, H and Takeuchi, S. Anomalous Weight Reduction on a Gyroscope's Right Rotations around the Vertical Axis on the Earth, *Phys.Rev.Lett.*, 63 (25), 1989: 2701- 2704.

[7] Hayasaka, H., Parity Breaking of Gravity and Generation of Antigravity due to the de Rham Cohomology Effect on Object's Spinning, Selected Papers of 3-rd Inter. Conf. on Problems of Space, Time, Gravitation, May 22-27, 1994 (held by Russian Academy of Science), St-Petersburg, Russia.

[8] Hayaska, H., Tanaka, H., Hasida, T., Chubachi, A., and Sugiyama, T. Possibility for the Existence of Anti-Gravity: Evidence from a Free-Fall Experiment Using a Spinning Gyro, *Speculations of Science and Technology*, 20, 1997: 173.

[9] Hayasaka H., Minami Y., Repulsive Generation Due to Topological Effect of Circulating Magnetic Fluids, STAIF-99, edited by Mohamed S. El-Genk, CP458, ,Jan.31-Feb.4, 1999: 1040-1050, Albuquerque, NM, USA.

[10] Ivanov B.V., Strong gravitational force induced by static electromagnetic fields, Available from: arXiv: gr-qc/047048 v1 (2004)

[11] Ivanov B.V., On the gravitational field induced by static electromagnetic sources, Available from: arXiv: gr-qc/0502047 v1 (2005)

[12] Ivanov B.V., Weyl electrovacuum solutions and gauge invariance, Available from: arXiv: gr-qc/0507082 v1 (2005)

[13] INTEL, Toward Flight without Stress or Strain … or Weight, *Interavia*, XI, 1956:373-374.

[14] Lahoz D.G., Graham G.M., Measurement of forces related to electromagnetic momentum in media at low frequencies, *Can. J. Phys.*, 57, 1979: 667-676.

[15] Lahoz D.G., Graham G.M., Observation of Electromagnetic Angular Momentum within Magnetite, *Physical Review Letters*, 42 (17), 1979: 1137-1140.

[16] Li N., Torr D.G., Gravitational effects on the magnetic attenuation of superconductors, *Phys.Rev. B*, 46 (9), 1992: 5489-5495.

[17] Li N., Noever D.,.Robertson T., Koczor R., Brantley W., Static test for a gravitational force coupled to type II YBCO superconductor, *Physica C*, 281,1997: 260-267.

[18] Martins A.A., Pinheiro M.J., Fluidic electrodynamics: Approach to electromagnetic propulsion, Space, Propulsion & Energy Sciences International Forum-SPESIF-2009, AIP, 2009: 216-224.

[19] Martins A.A., Pinheiro M.J., Fluidic electrodynamics: Approach to electromagnetic propulsion, *Physics of Fluids*, 21, 2009: 09713-1-7.

[20] Minami,Y., A Superstring-Based Field Propulsion Concept, *JBIS*, 57, 2004: 216-224.

[21] Natsuume M., Can String Theory Solve the Puzzles of Black Holes? , *Journal of the Physical Society of Japan*, 54, 178, March 1999.

[22] Pinheiro M.J., Electromagnetoroid structures in propulsion and astrophysics, AIP Conf. Proc. 1208 (2010) 186-191.

[23] Podkletnov E. Nieminen ,R, A possible gravitational force shielding by bulk $YBa_2Cu_3O_{7-x}$ superconductor, *Physica C*, 203, 1992: 441-444.

[24] Robertson G.A., Murad P.A., Davis E, New Frontiers in Space Propulsion Sciences, *Energy Conversion and Management*, 49, 2008: 436-452.

[25] Slepian J., Electromagnetic Space-Ship, *Electrical Engineering*, 68, 1949: 145-149.

[26] Torr D.G., Li N., Graviotoelectric-Electric Coupling via Superconductivity, *Foundations of Physics*, 6 (4) ,1993: 371-383.

[27] Wilson J., NASA's Antigravity Machine, *Popular Mechanics*, Dec 1977: 44-45.

CONCLUSIONS

[1] Cole D.C., Puthoff H., Extracting energy and heat from the vacuum, *Phys. Rev. E*, 48 (2), 1993: 1562-1565.

[2] Cole, D.C, Energy and Thermodynamic Considerations Involving Electromagnetic Zero-Point Radiation, Proceedings of Space Technology and Applications International Forum-1999, American Institute of Physics ,1999: 960-966.

[3] Davis E.W., Teofilo V.L., Haisch B., Puthoff H.E., Nickisch L.J., Rueda A., Cole D.C., Review of Experimental Concepts for Studying the Quantum Vacuum Field, Space Technology and Applications International Forum-STAIF 2006, AIP, 2006: 1390-1401.

[4] Minami Y., Liquid Metal MHD Power Generation System Using Antiproton Annihilation Reactor, in Survey Report of Research Committee on Functional New Material "Prometheus in Space", pp.136-150,March 1993, JAPAN SPACE UTILIZATION PROMOTION CENTER.

[5] Minami Y., Possibility of Space Drive Propulsion, paper IAA-94-IAA.4.1.658, presented at 45th IAF Congress 1994.

BIBLIOGRAPHIES

CHAPTER 2

1. Ball K.J and G. F. Osborne G.F., Space Vehicle Dynamics, Clarendon Press, Oxford, 1967.
2. Jahn R.G., Physics of Electric Propulsion, McGraw-Hill, New York, 1968.
3. Wiech,Jr. R.E. and Strauss R.F., Fundamentals of Rocket Propulsion, Reinhold Publishing, New York, 1960.

CHAPTER 3

1. De Felice F. and. Clarke C.J.S., Relativity on curved manifolds, Cambridge University Press, Cambridge, 1990.
2. Foster J. and Nightingale J.D., A Short Course in General Relativity, Springer-Verlag, New York, 1994.
3. Martin J.L., General Relativity : A first course for physicists, Prentice Hall, New York, 1996.
4. Milonni P.W., The Quantum Vacuum: An Introduction to Quantum Electrodynamics, Academic Press, San Diego, 1994.
5. Misner C.W., Thorne K.S. and Wheeler J.A., Gravitation, Freeman, San Francisco, 1973.
6. Pauli W., Theory of Relativity, Dover Publications, Inc, New York, 1981.

CHAPTER 4

1. Borg S.F., Matrix-Tensor Methods In Continuum Mechanics, D. Van Nostrand Company, Inc., New York, 1963.
2. Fung Y. C., Classical and Computational Solid Mechanics, World Scientific Publishing Co. Pre. Ltd., 2001.
3. Flügge W., Tensor Analysis and Continuun Mechanics, Springer-Verlag, New York, 1972.
4. Gerry, C. C., Knight, P. L., Introductory Quantum Optics, Cambridge University Press, Cambridge, 2005: 1-194.
5. Hawking S., HAWKING ON THE BIG BANG AND BLACK HOLES (Advanced Series in Astrophysics and Cosmology-Vol.8), World Scientific, Singapore, 1993.
6. Hawking S. and Penrose R., The Nature of Space and Time, Princeton University Press, 1996.
7. Hughston L.P and Tod K.P., An Introduction to General Relativity, Cambridge University Press, Cambridge, 1990.
8. Kaku M., Quantum Field Theory, Oxford University Press, Inc., 1993.
9. Kane G., Modern Elementary Particle Physics, Addison-Wesley, New York, 1993.
10. Kolb E.W. and Turner M.S., The Early Universe, Addison-Wesley, New York, 1993.
11. Matsuoka, M., Quantum Optics, Shokabo, Tokyo Japan, 2000, pp. 86-169.
12. Milton K A, The Casimir Effect: Physical Manifestations of Zero-Point Energy, World Scientific, Singapore, 2001.
13. Moss I.G., Quantum Theory, Black Holes and inflation, John Wiley & Sons, 1996.
14. Peskin M.E. and Schroeder D.V., An introduction to Quantum Field Theory, Persus Books, 1995.
15. Timoshenko S. and Woinowsky-Krieger S., Theory Of Plates And Shells, McGraw-Hill Book ompany, New York, 1959.
16. Tolman R.C., Relativity Thermodynamics and Cosmology, Dover Books, New York, 1978.
17. Umezawa H., Advanced Field Theory, American Institute of Physics, 1993.
18. Vedral V., Modern Foundations of Quantum Optics, Imperial Collage Press, London, 2005: 131-169.
19. Walls D.F., Milburn, G. J., Quantum Optics, Springer-Verlag, Berlin-Heidelberg, 1994: 1-228.

CHAPTER 5

1. Hawking S., A Brief History of Times, Bantam Publishing Company, New York, 1988.

2. Hawking S., HAWKING ON BIGBANG AND BLACK HOLE, World Scientific, 1993.
3. Visser M., Lorentzian Wormholes - From Einstein to Hawking, AIP PRESS, Woodbury, New York, 1996.

CHAPTER 6

1. Bailin D. and Love A., Supersymmetric Gauge Field Theory and String Theory, Taylor & Francis, New York, 1994.
2. Cai R-G, Myung Y.S. and Ohta N., Bekenstein Bound., Holography and Brane Cosmology in Charged Black Hole Backgrounds, *Classical and Quantum Gravity*, 18, 2001: 5429-5440.
3. Callan Jr. C.G. and Maldacena J.M., D-brane approach to black hole quantum mechanics, *Nucl. Phys. B.* 472,1996: 591-608.
4. Green M.B, Schwarz J.H. and Witten E., Superstring theory Vol.1,2, Cambridge Unversity Press, Cambridge, 1988.
5. Horowitz G., The origin of Black Hole Entropy in String Theory, gr-qc/9604051.
6. Horowitz G. and Strominger A., Counting States of Near-Extremal Black Holes,
1. *Phys. Rev. Lett.* 77,1996: 2368-2371.
7. Johnson C.V., D-Branes, Cambridge Monographs On Mathematical Physics, 2003.
8. Kaku M., Introduction to Superstrings and M-Theory 2nd ed, Springer, 1999.
9. Maldacena J.M., Black Holes in String Theory, hep-th/9607235.
10. Ohta N., Intersection Rules for Non-Extreme p-Branes, *Phys. Lett. B.* 403,1977: 218-224.
11. Polchinski J., String Theory, Vol.1,2, Cambridge University Press, Cambridge, 2005.
12. Peat F.D., Superstrings and Search for The Theory of Everything, CONTEMPORARY BOOKS, 1988.
13. Strominger A. and Vafa C., Microscopic Origin of the Bekenstein-Hawking Entropy, *Phys. Lett. B.*,379,1996: 99-104.
14. Verlinde E., On the Holographic Principle in a Radiation Dominated Universe, hep-th/0008140.
15. Zwiebach B., A First Course in String Theory, Canbridge University Press, 2004.

APPENDIX.A

MECHANICS OF SPACE

Expanding the concept of vector parallel displacement in Riemann space, the following equation has newly been obtained;

$$\omega_{\mu\nu} = R_{\mu\nu kl} dA^{kl} ,$$
(A.1)

where dA^{kl} is infinitesimal areal element.

According to the nature of Riemann curvature tensor $R_{\mu\nu kl}$, $\omega_{\mu\nu}$ indicates the rotation of displacement field. Eq.(A.1) indicates that a curved space produces the rotation of displacement field in the region of space. Now, the rotation tensor $\omega_{\mu\nu}$ and strain tensor e_{ij} satisfy the following differential equation in continuum mechanics.

$$\omega_{\mu\nu,j} = e_{\nu j,\mu} - e_{\mu j,\nu} .$$
(A.2)

This equation is true on the condition that the order of differential can be exchanged in a flat space. To expand above equation into a curved Riemann space, the equation shall be transformed to covariant differentiation and it is possible on the following condition: $\Gamma^\alpha_{j\nu} e_{\mu\alpha} = \Gamma^\alpha_{j\mu} e_{\nu\alpha}$.

Thus, we obtain

$$\omega_{\mu\nu:j} = e_{\nu j:\mu} - e_{\mu j:\nu} .$$
(A.3)

Here we use the usual notation ":" for covariant differentiation.

Eq.(A.3) indicates that the displacement gradient of rotation tensor corresponds to difference of the displacement gradient of strain tensor. Here, if we multiply both sides of Eq.(A.3) by a fourth order tensor related to the nature of space $E^{ij\mu\nu}$, we obtain formally

$$E^{ij\mu\nu}\omega_{\mu\nu:j} = E^{ij\mu\nu}(R_{\mu\nu kl}dA^{kl})_{:j} = E^{ij\mu\nu}R_{\mu\nu kl:j}dA^{kl} ,$$
(A.4)

and

$$E^{ij\mu\nu}e_{\nu j:\mu} - E^{ij\mu\nu}e_{\mu j:\nu} = (E^{ij\mu\nu}e_{\nu j})_{:\mu} - (E^{ij\mu\nu}e_{\mu j})_{:\nu} = \sigma^{i\mu}{}_{:\mu} - \sigma^{i\nu}{}_{:\nu} = \Delta\sigma^{ir}{}_{:r}$$
(A.5)

In the continuum mechanics, the relationship between stress tensor σ_{ij} and strain tensor e_{ml} is given by

$$\sigma^{ij} = E^{ijml}e_{ml} .$$
(A.6)

Furthermore, the relationship between body force F^i and stress tensor σ_{ij} is given by

$$F^i = \sigma^{ij}{}_{:j} ,$$
(A.7)

from the equilibrium conditions of continuum. Therefore, Eq.(A5) indicates the difference of body force ΔF^i. Accordingly, from Eqs(A.4) and (A.5), the change of body force ΔF^i becomes

$$\Delta F^i = E^{ij\mu\nu} R_{\mu\nu kl:j} dA^{kl} .$$ **(A.8)**

Here, we assumed that $E^{ij\mu\nu}$ is constant for covariant differentiation.

Eq.(A.8) indicates that the gradient of Riemann curvature tensor implying space curvature produces the body force as a space strain force. The space strain force is the body force.

APPENDIX.B

RELATIONSHIP TO SPECIAL RELATIVITY

Using Eq.(1.21) in Chap. 4: $ds'^2 - ds^2 = (g'_{ij} - g_{ij})dx^i dx^j = 2e_{ij} dx^i dx^j$, **(B.1)**

and adopting the following fundamental axiom, i.e., "The nature of flat space is identical independent of stationary system and uniform moving system and rectilinearly", we arrive to the following equation;

$$ds'^2 - ds^2 = (\eta'_{ij} - \eta_{ij})dx^i dx^j = 2e_{ij} dx^i dx^j = 0 \cdot dx^i dx^j .$$ **(B.2)**

Since the metric tensor of flat space is Minkowski metric η_{ij} , above axiom gives

$\eta'_{ij} = \eta_{ij}$. We arrive to the following two conditions from Eq.(B.2).

$$ds'^2 - ds^2 = 0 ,$$ **(B.3)**

and

$$e_{ij} = 0 .$$ **(B.4)**

From Eq.(B.3), we have

$$x'^2 + y'^2 + z'^2 - (c't')^2 = x^2 + y^2 + z^2 - (ct)^2 .$$ **(B.5)**

Then we can obtain the following equations from Eq.(B.5) in accordance with the usual calculation in Special Relativity.

$$x' = \gamma(x - \beta ct),\ y' = y,\ z' = z,\ c't' = \gamma(ct - \beta x) ,$$ **(B.6)**

where $\gamma = 1/\sqrt{1-\beta^2}$, c' is the speed of light in moving coordinate system and c is the speed of light in stationary coordinate system.

We must pay attention that we do not require the so-called "Principle of constancy of light velocity" at present.

From Eq.(B4) we have

$$e_{ij} = u^i_{\ ;i} = div\, u = 0 .$$ **(B.7)**

Since Eq.(B7) indicates that the divergence of displacement vector "u", i.e., volume strain is zero, there exists the following wave equation of space-time satisfying $div\, u = 0$, that is

$$\nabla^2 u^i - \frac{1}{c_t^2} \cdot u^i = 0 .$$ **(B.8)**

And also, this wave equation of motion requires the transversal wave "c_t".

By doing trial calculation of D'Alembertian $\nabla^2 - 1/c_t^2 \cdot \partial^2/\partial t^2$ of wave equation using the result of Eq.(B.8) for both coordinate systems, the wave velocity of both coordinate systems becomes identical, that is

$$c' = c.$$

(B.9)

And it is required that wave velocity of space is always constant.

Thus, we can get the principle of constancy of light velocity. Substituting the result of Eq.(B.9) into Eq.(B.6), we can obtain Lorentz transformation. As a result, the principle of constancy of light velocity is embedded in the space-time as a physical nature from the outset.

APPENDIX.C

CURVATURE AND ACCELERATION

A massive body causes the curvature of space-time around it and a free particle responds to it by moving along a geodesic in this region of space-time. The path of free particle is a geodesic in space-time and is given by the following equation;

$$\frac{d^2x^i}{d\tau^2} + \Gamma^i_{jk} \cdot \frac{dx^j}{d\tau} \cdot \frac{dx^k}{d\tau} = 0.$$

(C.1)

From Eq.(C.1), the acceleration of free particle is obtained by

$$\alpha^i = \frac{d^2x^i}{d\tau^2} = -\Gamma^i_{jk} \cdot \frac{dx^j}{d\tau} \cdot \frac{dx^k}{d\tau}.$$

(C.2)

Eq.(C.2) yields a more simple equation from the condition of linear approximation, that is, weak-field, quasi-static and slow motion (speed $v \ll$ speed of light c);

$$\alpha^i = -\sqrt{-g_{00}} \cdot c^2 \Gamma^i_{00}.$$

(C.3)

On the other hand, the major component of spatial curvature R^{00} in the weak field is given by

$$R^{00} \approx R_{00} = R^\mu_{0\mu0} = \partial_0\Gamma^\mu_{0\mu} - \partial_\mu\Gamma^\mu_{00} + \Gamma^\nu_{0\mu}\Gamma^\mu_{\nu0} - \Gamma^\nu_{00}\Gamma^\mu_{\nu\mu}.$$

(C.4)

In the nearly Cartesian coordinate system, the value of $\Gamma^\mu_{\nu\rho}$ are small, so we can neglect the last two terms in Eq.(C.4) and using the quasi-static condition we get

$$R^{00} \approx -\partial_\mu\Gamma^\mu_{00} = -\partial_i\Gamma^i_{00}.$$

(C.5)

From Eq.(C.5), we get formally

$$\Gamma^i_{00} = -\int R^{00}(x^i)dx^i.$$

(C.6)

Substituting Eq.(C.6) into Eq.(C.3), we obtain

$$\alpha^i = \sqrt{-g_{00}}\,c^2 \int R^{00}(x^i)\,dx^i \;.$$

$$(C.7)$$

APPENDIX.D

CURVATURE CONTROL BY MAGNETIC FIELD

Let us consider the electromagnetic energy tensor M^{ij}. In this case, the solution of metric tensor g_{ij} is found by

$$R^{ij} - \frac{1}{2}\cdot g^{ij}R = -\frac{8\pi G}{c^4}\cdot M^{ij}\;.$$

$$(D.1)$$

Eq.(D.1) determines the structure of space-time due to the electromagnetic energy.

Here, if we multiply both sides of Eq.(D.1) by g_{ij}, we obtain

$$g_{ij}\left(R^{ij} - \frac{1}{2}\cdot g^{ij}R\right) = g_{ij}R^{ij} - \frac{1}{2}\cdot g_{ij}g^{ij}R = R - \frac{1}{2}\cdot 4R = -R\;,$$

$$(D.2)$$

$$g_{ij}\left(\frac{-8\pi G}{c^4}\cdot M^{ij}\right) = -\frac{8\pi G}{c^4}\cdot g_{ij}M^{ij} = \frac{-8\pi G}{c^4}\cdot M_i^i = \frac{-8\pi G}{c^4}M\;.$$

$$(D.3)$$

The following equation is derived from Eqs.(D.2) and (D.3)

$$R = \frac{8\pi G}{c^4}\cdot M\;.$$

$$(D.4)$$

Substituting Eq.(D.4) into Eq.(D.1), we obtain

$$R^{ij} = -\frac{8\pi G}{c^4}\cdot M^{ij} + \frac{1}{2}\cdot g^{ij}R = -\frac{8\pi G}{c^4}\cdot\left(M^{ij} - \frac{1}{2}\cdot g^{ij}M\right).$$

$$(D.5)$$

Using antisymmetric tensor f_{ij} which denotes the magnitude of electromagnetic field, the electromagnetic energy tensor M^{ij} is represented as follows;

$$M^{ij} = -\frac{1}{\mu_0}\cdot\left(f^{i\rho}f_\rho^{\,j} - \frac{1}{4}\cdot g^{ij}f^{\alpha\beta}f_{\alpha\beta}\right),\quad f^{i\rho} = g^{i\alpha}g^{\rho\beta}f_{\alpha\beta}\;.$$

$$(D.6)$$

Therefore, for M, we have

$$\begin{aligned}
M = M_i^i = g_{ij}M^{ij} &= -\frac{1}{\mu_0}\cdot\left(g_{ij}f^{i\rho}f_\rho^{\,j} - \frac{1}{4}\cdot g_{ij}g^{ij}f^{\alpha\beta}f_{\alpha\beta}\right)\\
&= -\frac{1}{\mu_0}\cdot\left(f^{i\rho}f_{i\rho} - \frac{1}{4}\cdot 4f^{\alpha\beta}f_{\alpha\beta}\right) = -\frac{1}{\mu_0}\cdot\left(f^{i\rho}f_{i\rho} - f^{i\rho}f_{i\rho}\right) = 0
\end{aligned}$$

$$(D.7)$$

Accordingly, substituting $M=0$ into Eq.(D.5), we get

$$R^{ij} = -\frac{8\pi G}{c^4}\cdot M^{ij}\;.$$

$$(D.8)$$

Although Ricci tensor R^{ij} has 10 independent components, the major component is the case of $i=j=0$, i.e., R^{00}. Therefore, Eq.(D.8) becomes

$$R^{00} = -\frac{8\pi G}{c^4} \cdot M^{00}.$$

(D.9)

On the other hand, 6 components of antisymmetric tensor $f_{ij} = -f_{ji}$ are given by electric field E and magnetic field B from the relation to Maxwell's field equations

$$f_{10} = -f_{01} = \frac{1}{c} \cdot E_x, f_{20} = -f_{02} = \frac{1}{c} \cdot E_y, f_{30} = -f_{03} = \frac{1}{c} E_z$$
$$f_{12} = -f_{21} = B_z, f_{23} = -f_{32} = B_x, f_{31} = -f_{13} = B_y$$
$$f_{00} = f_{11} = f_{22} = f_{33} = 0$$

(D.10)

Substituting Eq.(D.10) into Eq.(D.6), we have

$$M^{00} = -\left(\frac{1}{2} \cdot \varepsilon_0 E^2 + \frac{1}{2\mu_0} \cdot B^2\right) \approx -\frac{1}{2\mu_0} \cdot B^2.$$

(D.11)

Finally, from Eqs.(D.9) and (D.11), we have

$$R^{00} = \frac{4\pi G}{\mu_0 c^4} \cdot B^2 = 8.2 \times 10^{-38} \cdot B^2 \quad (B \text{ in Tesla}),$$

(D.12)

where we let $\mu_0 = 4\pi \times 10^{-7} (H/m)$, $\varepsilon_0 = 1/(36\pi) \times 10^{-9} (F/m)$, $c = 3 \times 10^8 (m/s)$,

$G = 6.672 \times 10^{-11} (N \cdot m^2/kg^2)$, B is a magnetic field in Tesla and R^{00} is a major component of special curvature $(1/m^2)$.

The relationship between curvature and magnetic field was derived by Minami and introduced it in 16th International Symposium on Space Technology and Science (1988).

Eq.(D12) is derived from general method.

On the other hand, Levi-Civita also investigated the gravitational field produced by a homogeneous electric or magnetic field, which was expressed by Pauli (Pauli, 1981). If x^3 is taken in the direction of a magnetic field of intensity F (Gauss unit), the square of the line element is of the form;

$$ds^2 = (dx^1)^2 + (dx^2)^2 + (dx^3)^2 + \frac{(x^1 dx^1 + x^2 dx^2)^2}{a^2 - r^2}$$
$$-\left[c_1 \exp(x^3/a) + c_2 \exp(-x^3/a)\right]^2 (dx^4)^2,$$

(D.13)

where $r = \sqrt{(x^1)^2 + (x^2)^2}$, c_1 and c_2 are constants, $a = \frac{c^2}{\sqrt{kF}}$, k is Newtonian gravitational constant(G) and $x^1...x^4$ are Cartesian coordinates ($x^1...x^3$=space, x^4=ct) with orthographic projection.

The space is cylindrically symmetric about the direction of the field and on each plane perpendicular to the field direction the same geometry holds as in Euclidean space on a sphere of radius a, that is, the radius of curvature a is given by

$$a = \frac{c^2}{\sqrt{kF}} \; . \tag{D.14}$$

Since the relation of between magnetic field B in SI units and magnetic field F in CGS Gauss units are described as follows: $B\sqrt{\frac{4\pi}{\mu_0}} \Leftrightarrow F$, then the radius of curvature "a" in Eq.(D14) is expressed in SI units as the following (changing symbol, $k \rightarrow G, F \rightarrow B$):

$$a = \frac{c^2}{\sqrt{GF}} = \frac{c^2}{\sqrt{G} \cdot B \sqrt{\frac{4\pi}{\mu_0}}} \approx 3.484 \times 10^{18} \frac{1}{B} \quad meters \; .$$

While, scalar curvature is represented by $R^{00} \approx R = \frac{1}{a^2} = \frac{GB^2 \frac{4\pi}{\mu_0}}{c^4} = \frac{4\pi G}{\mu_0 c^4} B^2$,

which coincides with (D.12).

APPENDIX.E

ACCELERATION INDUCED BY DE SITTER SOLUTION

The most general form of gravitational field equation which includes cosmological constant is given by

$$R^{ij} - \frac{1}{2} \cdot g^{ij} R = -\frac{8\pi G}{c^4} T^{ij} - \Lambda g^{ij}, \tag{E.1}$$

where R^{ij} is the Ricci tensor, R is the scalar curvature, G is the gravitational constant, c is the velocity of light, T^{ij} is the energy momentum tensor, and Λ is the cosmological constant.

It is simple to see that a cosmological term Λg^{ij} is equivalent to an additional form of energy momentum tensor. The cosmological term is identical to the energy momentum tensor associated with the vacuum.

From the expansion and contraction of Eq.(E1), we get the following respectively;

$$R^{00} = 4\Lambda \;, and \; R^{ij} = \Lambda g^{ij} \; . \tag{E.2}$$

Eq.(E2) means that the cosmological constant Λ generates the major component of curvature of space. Therefore, the curvature of space is identical with cosmological constant.

Now, concerning the de Sitter cosmological model with non-zero vacuum energy (i.e. cosmological constant), the de Sitter line element is written as

$$ds^2 = -(1 - \frac{1}{3}\Lambda r^2)c^2 dt^2 + \frac{1}{1 - \frac{1}{3}\Lambda r^2}dr^2 + r^2(d\theta^2 + \sin^2\theta d\varphi^2) \; . \tag{E.3}$$

The metrics are given by

$$g_{00} = -(1 - 1/3 \cdot \Lambda r^2), \; g_{11} = g_{22} = 1, \; g_{33} = 1/(1 - 1/3 \cdot \Lambda r^2),$$
$$and \; other \; g_{ij} = 0 \tag{E.4}$$

As mentioned previously, we have

$$\alpha = \sqrt{-g_{00}}\, c^2 \Gamma_{00}^3, \quad \Gamma_{00}^3 = \frac{-g_{00,3}}{2g_{33}}. \tag{E.5}$$

The acceleration α of de Sitter solution can be obtained by combining Eq.(E.5) with Eq.(E.4);

$$\alpha = \frac{1}{3} c^2 \Lambda r \quad (1 > 1/3 \cdot \Lambda r^2). \tag{E.6}$$

The acceleration induced by cosmological constant is proportional to the distance "r" from the generative source, i.e. engine system.

According to the gauge theories, the physical space as a vacuum is filled with a spin-zero scalar field, called a Higgs field. The vacuum energy fluctuates in proportion to the fluctuation of Higgs field. The vacuum potential $V(\varphi)$ is given by the vacuum expectation value φ of Higgs field, and we get the minimum of the Higgs potential $V_0(\varphi)$ as follows:

$$V_0(\phi) = \frac{\lambda}{4} \phi_0^{\,4}, \tag{E.7}$$

where λ is the constant, φ_0 is the non-zero vacuum expectation value of Higgs field.

Since the vacuum potential $V_0(\varphi)$ shall be invariant under the Lorentz transformation, the energy momentum tensor of vacuum $T^{ij}_{\ vac}$ is written in the form as

$$T^{ij}_{\ vac} = V_0(\phi) g^{ij}. \tag{E.8}$$

The energy momentum tensor of vacuum exerts the same action as that of the cosmological term. It should be noted that $T^{ij}_{\ vac}$ is not energy momentum tensor of matter but the vacuum itself. From Eq.(E.1) and Eq.(E.8), as its metric source, $8\pi G / c^4 \cdot T^{ij}_{\ vac} = 8\pi G / c^4 \cdot V_0(\phi) g^{ij} = \Lambda g^{ij}$, then we get

$$\Lambda = \frac{8\pi G}{c^4} V_0(\phi) = 2.1 \times 10^{-43} V_0(\phi). \tag{E.9}$$

In general, since the potential from its source is inversely proportional to the distance "r" from the potential source, assuming that the vacuum potential $V_0(\varphi)$ in Eq.(E.7) is the energy source, the potential at distance "r" apart from its energy source is written in the form $V_0(\varphi) \Rightarrow \frac{V_0(\varphi)}{r} = \frac{\lambda}{4r} \varphi_0^4$.

Combining Eq.(E.9) with above equation yields $\Lambda = 2\pi G \varphi_0^4 / c^4 r$.

Substituting this equation into Eq.(E.6), we finally get

$$\alpha = \frac{2\pi G \lambda}{3c^2} \phi_0^{\,4} = 1.6 \times 10^{-27} \lambda \phi_0^{\,4}, \tag{E.10}$$

where G is a gravitational constant, c is a velocity of light, λ is an arbitrary Higgs self-coupling in the Higgs potential (λ is not known and is not determined by a gauge principle, presumably $\lambda \sim 1/10$), and φ_0 is a non-zero vacuum expectation value of Higgs field.

APPENDIX.F

PROPERTIES OF HYPER-SPACE

Let us put x^1, x^2, x^3 for x, y, z and x^0 for *ct*. In Minkowski space, the distance(s) are given by

$$S_{RS}^2 = \eta_{ij}x^i x^j = -(x^0)^2 + (x^1)^2 + (x^2)^2 + (x^3)^2 = -(ct)^2 + x^2 + y^2 + z^2 , \tag{F.1}$$

where η_{ij} is Minkowski metric, and c is the speed of light.

Eq.(F1) indicates the properties of the actual physical space limited by Special Relativity. From Eq.(F1), Minkowski metric, i.e. real-space metric is shown as follows:

$$\eta_{ij(RS)} = \begin{bmatrix} -1 & 0 & 0 & 0 \\ 0 & +1 & 0 & 0 \\ 0 & 0 & +1 & 0 \\ 0 & 0 & 0 & +1 \end{bmatrix}, \ \eta_{00}=-1, \ \eta_{11}=\eta_{22}=\eta_{33}=+1, \ and \ other \ \eta_{ij}=0 , \tag{F.2}$$

The properties of space are determined by the metric tensor, which defines the distance between two points.

Here, as a hypothesis, we demand an invariant distance for the time component of Minkowski metric reversal, i.e., $\eta_{ij} \to \eta_{ij}, \eta_{00} \to -\eta_{00}$.

This hypothesis gives the minimum distance between two kinds of space-time, that is, the properties of Hyper-Space are close to that of Minkowski space.

Namely, regarding the following equation;

$$(x^0)^2 \to -(x^0)^2 \left[= (ct)^2 \to -(ct)^2 \right], \textbf{i.e.,} \ \eta_{00}=-1 \to \eta_{00}=+1 ,$$

we require the following equation.

$$S_{HS}^2 = S_{RS}^2 , \tag{F.3}$$

where suf."HS" denotes Hyper-Space and suf."RS" denotes Real-Space.

From the above hypothesis, the metric of Hyper-Space becomes as follows:

$$\eta_{ij(HS)} = \begin{bmatrix} +1 & 0 & 0 & 0 \\ 0 & +1 & 0 & 0 \\ 0 & 0 & +1 & 0 \\ 0 & 0 & 0 & +1 \end{bmatrix}, \ \eta_{00}=+1, \ \eta_{11}=\eta_{22}=\eta_{33}=+1, \ and \ other \ \eta_{ij}=0 . \tag{F.4}$$

Therefore, in Hyper-Space, the distance is given by

$$S_{HS}^2 = \eta_{ij}x^i x^j = (x^0)^2 + (x^1)^2 + (x^2)^2 + (x^3)^2 = (ct)^2 + x^2 + y^2 + z^2 . \tag{F.5}$$

Accordingly, Hyper-Space shows the properties of Euclidean space. Therefore, the imaginary time ($x^0 = ict$; $i^2=-1$) as the component of time coordinate is required as a necessary result in Hyper-Space. The time "t" in Real-Space is changed to "it" in Hyper-Space. Here, "i" denotes the imaginary unit. However, the components of space coordinates(x,y,z) are the same real numbers as Real-Space.

By substituting the imaginary time "it" into Eq,(F.5), we obtain Eq.(F.1), so that the invariance of distance is satisfied.

APPENDIX.G

LORENTZ TRANSFORMATION OF HYPER-SPACE

The Lorentz transformation of Hyper-Space corresponding to that of Real-Space is found. Since the components of space coordinates(x,y,z) do not change between Real-Space and Hyper-Space, the velocity in Hyper-Space can be obtained by changing $t \rightarrow it$;

$$V = dx / dt \rightarrow dx / d(it) = dx / idt = V / i = -iV \text{ .} \tag{G.1}$$

The velocity becomes the imaginary velocity in Hyper-Space. Substituting "$t \rightarrow it$,

$V \rightarrow -iV$ " into the Lorentz transformation equations of Minkowski space formally gives

< Hyper-Space Lorentz transformation >

$$x' = (x - Vt) / [1 + (V / c)^2]^{1/2}, \quad t' = (t + Vx / c^2) / [1 + (V / c)^2]^{1/2},$$
$$\Delta t' = \Delta t [1 + (V / c)^2]^{1/2}, \quad L' = \Delta L [1 + (V / c)^2]^{1/2} \tag{G.2}$$

This result agrees with the results of detailed calculation. As a reference, the Lorentz transformation equations of Minkowski space, i.e. of Special Relativity, are shown below;

< Real-Space Lorentz transformation: Special Relativity >

$$x' = (x - Vt) / [1 - (V / c)^2]^{1/2}, \quad t' = (t - Vx / c^2) / [1 - (V / c)^2]^{1/2},$$
$$\Delta t' = \Delta t [1 - (V / c)^2]^{1/2}, \quad \Delta L' = \Delta L [1 - (V / c)^2]^{1/2} \tag{G.3}$$

The main difference is that the Lorentz-Fitz Gerald contraction factor $[1 - (V / c)^2]^{1/2}$ is changed to $[1 + (V / c)^2]^{1/2}$.

Now consider the navigation with the help of both Lorentz transformations, especially the Lorentz contraction of time.

Figure G.1 shows a transition of starship from Real-Space to Hyper-Space. In Fig.G.1, region I stands for Real-Space (Minkowski space).

Consider two inertial coordinate systems, S and S'. S' moves relatively to S at the constant velocity of starship (V_S) along the x-axis. S' stands for the coordinate system of the starship and S stands for the rest coordinate system $(V_S=0)$ on the earth. Δt_{ERS} is the time of an observer on the earth, i.e. earth time, and $\Delta t'_{RS}$ is the time shown by a clock in the starship, i.e. starship time.

Region II stands for Hyper-Space(Euclidean space). S' moves relatively to S at the constant velocity of starship (V_S) along the x-axis. S' stands for the coordinate system of the starship in Hyper-Space and S stands for the rest coordinate system $(V_S=0)$ in Hyper-Space. Δt_{EHS} is the time of an observer on the earth in Hyper-Space, i.e. the equivalent earth time, and $\Delta t'_{HS}$ is the time shown by a clock in the starship in Hyper-Space, i.e. the starship time. Now, the suffix."HS" denotes Hyper-Space and the suffix."RS" denotes Real-Space.

Figure.G1 also shows a linear mapping *f:RS(Real-Space)→HS(Hyper-Space)*, that is, from a flat Minkowski space-time manifold to a flat imaginary space-time manifold. It is assumed that space is an infinite continuum. There exists a 1-1 map *f:RS→HS, x^i|→f(x^i)* and a 1-1 inverse map *f^{-1}:HS→RS, f(x^i)|→x^i*. The mapping is a bijection. These transformations will be local and smooth.

Now suppose that a starship accelerates in Real-Space and achieves a quasi-light velocity (V_S~c) and plunges into Hyper-Space by some new technical methods.

In Real-Space, from Eq.(G3), we have

$$\Delta t'_{RS} = \Delta t_{ERS}[1-(V_S/c)^2]^{1/2} .$$ **(G.4)**

Eq.(G.4) is the so-called Lorentz contraction of time derived from Special Relativity. In Hyper-Space, the starship keeps the same velocity as the quasi-light velocity (V_S~c) just before plunging into Hyper-Space, i.e. $V_{S(HS)}=V_{S(RS)}$. Therefore, from Eq.(G.2), we have

$$\Delta t'_{HS} = \Delta t_{EHS}[1+(V_S/c)^2]^{1/2} .$$ **(G.5)**

From Fig.G1, after plunging into Hyper-Space, the starship keeps the quasi-light velocity and takes the S' coordinates. The elapsed time in the starship will be continuous. Considering the continuity of starship time between Real-Space and Hyper-Space, we get

$$\Delta t'_{RS} = \Delta t'_{HS} .$$ **(G.6)**

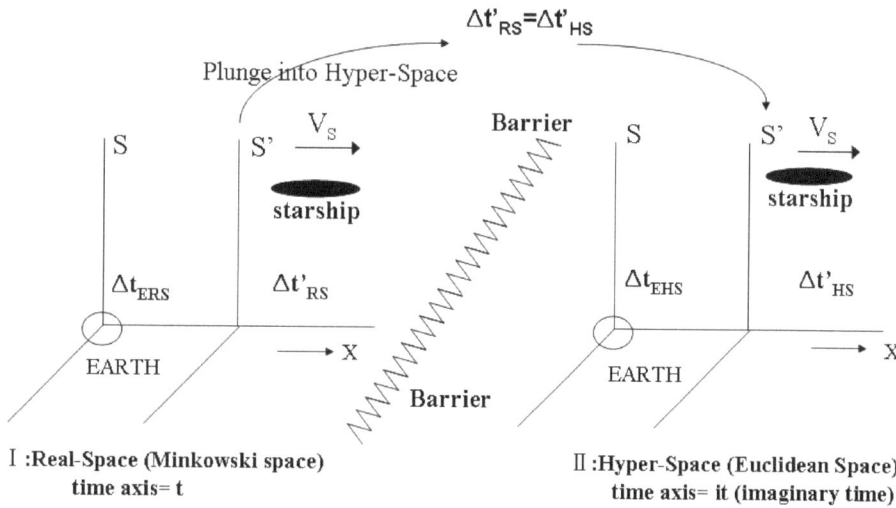

Figure G1: Transition from Real-Space to Hyper-Space

Now Eq.(G6) gives, from Eqs.(G.4) and (G.5), we have

$$\Delta t_{ERS} = \Delta t_{EHS} \left(\frac{[1+(V_S/c)^2]^{1/2}}{[1-(V_S/c)^2]^{1/2}} \right) .$$ **(G.7)**

Eq.(G.7) is the time transformation equation of earth time between Real-Space and Hyper-Space. From Eq.(G.7), when V_S=0, we get

$$\Delta t_{ERS} = \Delta t_{EHS} .$$ **(G.8)**

Namely, in the reference frame at rest, the elapsed time on the earth coincides with both Real-Space and Hyper-Space. However, as the velocity of starship approaches the velocity of light, the earth time between Real-Space and Hyper-Space becomes dissociated on a large scale

Since an observer on the earth looks at the starship going at $V_S \sim c$ and loses sight of it as it plunges into Hyper-Space, it is observed that the starship keeps the same velocity and moves during the elapsed time Δt_{ERS} (at $V_S \sim c$) observed from the earth. Therefore, the range of starship of an observer on the earth is given by

$$L = V_S \Delta t_{ERS} \approx c \Delta t_{ERS} .$$
(G.9)

For instance, in the case of $V_S = 0.999999999c$, from Eqs.(G.7) and (G.5), we get

$$\Delta t_{ERS} = \Delta t_{EHS} \times 31622, \quad \Delta t'_{HS} = \Delta t_{EHS} \times 1.4 .$$
(G.10)

One second in Hyper-Space corresponds to 31,622 seconds in Real-Space. Similarly, one hour in Hyper-Space corresponds to 31,622 hours (3.6 years) in Real-Space.

While the starship takes a flight for 100 hours ($\Delta t'_{HS} = 100hr;$ $V_S = 0.999999999c$) shown by a clock in the starship in Hyper-Space, 70 hours ($\Delta t_{EHS} = 70hr; V_S = 0$) have elapsed on the earth in Hyper-Space. Since this elapsed time on the earth in Hyper-Space is in the reference frame at rest, the time elapsed in it is the same as the time elapsed on the earth in Real-Space ($[\Delta t_{EHS}; V_S = 0] = [\Delta t_{ERS}; V_S = 0] = 70hr$). Therefore, there is not much difference between the elapsed time (70 hours) of an observer on the earth in Real-Space and the elapsed time (100 hours) of starship during Hyper-Space navigation. However, this elapsed time of 70 hours ($\Delta t_{EHS} = 70hr; V_S = 0$) on the earth in Hyper-Space becomes the elapsed time of 253 years ($\Delta t_{ERS} = 70 \times 31,622 = 2,213,540hr; V_S = 0.999999999c$) on the earth in Real-Space, because the starship flies at the velocity of $0.999999999c$. These 253 years represents the flight time of starship observed from the earth in Real-Space.

Therefore, by plunging into Hyper-Space having the properties of imaginary time, from Eq.(G.9), the starship at a quasi-light velocity can substantially move a distance of approximately 253 light years. In this way, the starship at a quasi-light velocity can travel to the stars 253 light years away from us in just 100 hours.

The above numerical estimation depends on the velocity of starship. For instance, in the case of $V_S = 0.99999c$, we get

$$\Delta t_{ERS} = \Delta t_{EHS} \times 316, \quad \Delta t'_{HS} = \Delta t_{EHS} \times 1.4 .$$
(G.11)

On the contrary, in the case of $V_S = 0.999 \ldots \ldots 999c$, a gap between Δt_{ERS} and Δt_{EHS} rapidly increases. That depends on how the starship can be accelerated to nearly the velocity of light.

APPENDIX.H

METHODS OF HYPER-SPACE NAVIGATION

Here let us consider the concrete method of plunging into Hyper-Space and returning back to Real-Space.

H.1 MANY-PARTICLE SYSTEMS FOR SPACESHIP

A plunging into Hyper-Space from Real-Space can be performed every where in Real-Space, whenever the technical condition of spaceship is ready. Namely, the spaceship can plunge into Hyper-Space at any time without restriction of navigation course. This implies that Real-Space always coexists with Hyper-Space as a parallel space. A factor which isolates Real-Space from Hyper-Space is a usual real time and imaginary time. Such a state may be analogous to a state of de Sitter space. In general, when two kinds of diverse phase space coexist or adjoin, most naturally a potential barrier shall exist to isolate these two kinds of phase space. Therefore, the spaceship shall overcome the potential barrier. Next, we consider the value of above-stated potential barrier.

When the spaceship reaches the velocity of light, a space as a continuum reaches its limits, i.e. a fracture point and begins to crack. To fracture a space, the spaceship shall give its kinetic energy to the space as an external force. If this energy exceeds the crack energy of space, then the crack begins.

Hence, the space is to be fractured rapidly by this crack growth rate. However, this fracture is localized.

The energy (E_k) provided by spaceship is

$$E_K = 1/2 \cdot Mc^2 \, , \tag{H.1}$$

where M is the mass of spaceship and c is the velocity of light.

The value of E_k varies with mass. However, since the value of potential barrier shall be constant, some standard constant value of mass is required.

Here, we adopt the Planck mass m_{PL} which is the maximum mass in elementary particle. The Planck mass is given by

$$m_{PL} = (\hbar c / G)^{1/2} = 2.2 \times 10^{-8} kg \, , \tag{H.2}$$

where $\hbar$ equals to the Planck constant divided by 2π, c is the velocity of light and G is the gravitational constant.

The Planck mass consists of only fundamental constant and plays a significant role for the unification of all interactions.

In addition to Planck mass, we may define a Planck energy E_{PL} to be

$$E_{PL} = m_{PL}c^2 = (\hbar c^5 / G)^{1/2} = 1.9 \times 10^9 J \, . \tag{H.3}$$

In addition, the Planck length L_{PL} and Planck time t_{PL} are given respectively by:

$$L_{PL} = (G\hbar / c^3)^{1/2} = 1.6 \times 10^{-35} m \, , \tag{H.4}$$

$$t_{PL} = (G\hbar / c^5)^{1/2} = 5.4 \times 10^{-44} s \, . \tag{H.5}$$

Above Planck length and Planck time are the shortest length and time which have physical meaning in quantum theory. Furthermore, these values, i.e. m_{PL}, E_{PL}, L_{PL},

t_{PL} are essential constants which play a significant role in quantum cosmology.

From above, the energy of potential barrier V_{R-H} is given by

$$V_{R-H} = 1/2 \cdot m_{PL}c^2 = 9.8 \times 10^8 J \, . \tag{H.6}$$

Let us suppose that the spaceship of mass M is formed a fine-grained structure of N- Planck masses to be

$$M = N \cdot m_{PL} \, . \tag{H.7}$$

The spaceship forms a fine-grained structure and shall maintain the shape as many-particle systems to recreate the structure of spaceship that existed in the initial stage. It is necessary to subdivide the spaceship into the size of mass that has recreated the initial structure. Because it is impossible to recreate the spaceship if it is subdivided in the size of atoms or molecules. Therefore, the spaceship is composed of Planck mass of N particles. It is necessary for the spaceship to be formed by a fine-grained structure in order to treat the spaceship as a many-particle systems by some technical methods. Thus, we can apply a quantum tunneling effect to the spaceship.

From above, the kinetic energy of spaceship is represented by

$$E_K = 1/2 \cdot Mc^2 = 1/2 \cdot (m_{PL} \cdot N)c^2 = (1/2 \cdot m_{PL}c^2) \cdot N .$$ **(H.8)**

Accordingly, the spaceship of mass M can be transformed to Planck mass of N.

From above discussion, the potential barrier shall be constant for all the massive body, which can be obtained by Eq.(H.6).

H.2 PENETRATION OF POTENTIAL BARRIER BY QUANTUM TUNNELING

Let us suppose that the thickness of potential barrier is a Planck length.

This assumption is something like the potential thickness of de Sitter cosmological model. The Planck length is considered as a fundamental constant of space-time. By the method of a fine-grained structure technology, the mass of spaceship is subdivided into the Planck mass of N particles. Even if the energy of each particle is less than the potential barrier, the particle can tunnel through the barrier by the quantum tunneling effect. By quantum tunneling effect, the spaceship as a many-particle systems can plunge into Hyper-Space without fracture of space, even if its velocity is less than the velocity of light ($V_S < C$). In the case of the energy of particle less than potential barrier ($E < V$), the transmissivity T is given by

$$T = [1 + V^2 \sinh^2 \alpha d / 4E(V - E)]^{-1} , \quad \alpha = [2m(V - E)]^{1/2} / \hbar ,$$ **(H.9)**

where m(kg) is the mass of a particle, $V(J)$ is the height of the potential barrier, $E(J)$ is the energy of a particle and d(m) is the thickness of the potential barrier.

If $\alpha d < 1$, the following approximate equation is obtained

$$T = [1 + V^2 m d^2 / 2E\hbar^2]^{-1} .$$ **(H.10)**

If the shape of potential barrier is not square potential but like Gaussian shape, we get the following by WKB approximation;

$$T = \exp[-2\int_a^b \{2m(V(x) - E)\}^{1/2} / \hbar \cdot dx] \approx \exp[-2\alpha d] .$$ **(H.11)**

With the help of Eqs.(H.9),(H.10) and (H.11), let us estimate the transmissivity.

Substituting the following values for Eqs.(H.9)and (H.10), then

$$V = 1/2 \cdot m_{PL}c^2 = 9.9 \times 10^8 J, \ d = L_{PL} = 1.6 \times 10^{-35} m, \ V_S = 0.999999999 \, c,$$
$$E = 1/2 \cdot m_{PL}V_S^2 = 9.89999998 \times 10^8 J, \ V - E \approx 1.98J,$$

and we get $\alpha d = 4.5 \times 10^{-5} < 1$ and hence we obtain $T = 0.8$.

If we use Eq.(H.11), the result is

$$T = e^{-2\alpha d} = 0.99991 \approx 1.$$

In the case of WKB approximation, which the potential energy changes slowly as a function of position, the spaceship composed of many-particle systems can tunnel through this potential barrier and plunge into Hyper-Space.

And also, in the case of a square potential, almost all the particles can tunnel through the potential barrier. By doing repetition, all particles can tunnel through this potential barrier.

As can be seen from above, the transmissivity depends on the mass. If the mass $m=10m_{PL}, 100m_{PL}, 1/10 \cdot m_{PL}, 1/100 \cdot m_{PL}$, the transmissivity T becomes $T=0.28$, $T=0.038$, $T=0.975$ and $T=0.9975$ respectively, using Eq.(H.9) or (H.10).

Therefore, the large mass can not tunnel through the potential barrier. Anyway, the quantum tunneling effect has the advantage of plunging into Hyper-Space without the fracture of space that is attained near the velocity of light.

H.3 NAVIGATION SCENARIO BETWEEN REAL-SPACE AND HYPER-SPACE

Figure.H1 shows the navigation scenario of spaceship passing through Hyper-Space region. Although the spaceship is a massive body of M at a certain time, the spaceship is formed of a fine-grained structure as a many-particle system of $m_{PL} \times N$. The wave function of spaceship is required at this moment. To do so, the spaceship turns on a fine-grained structure technology. After that, the spaceship composed of many-particle systems plunges into Hyper-Space.

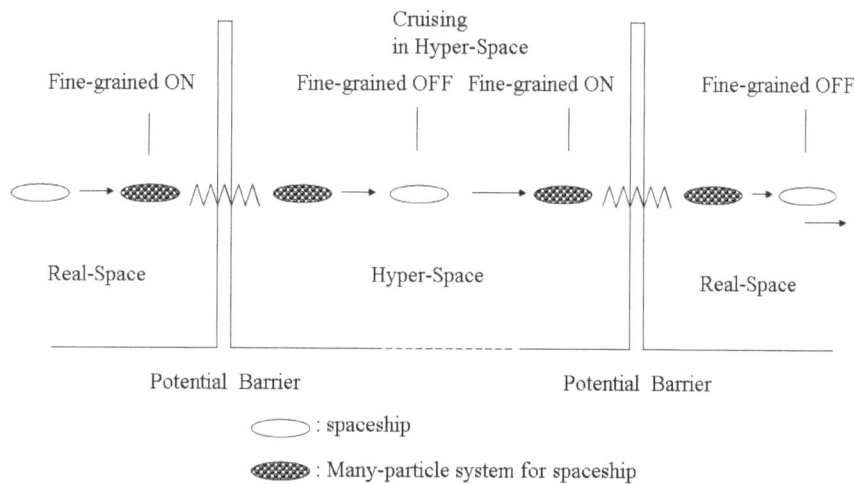

Figure H1: Navigation scenario of Spaceship

Then, the spaceship turns off a fine-grained structure technology and continues to travel in Hyper-Space. In order to jump out from Hyper-Space, the spaceship turns on a fine-grained structure technology again and plunges into Real-Space by quantum tunneling effect. After that, the spaceship turns off a fine-grained structure technology again, then decelerates and continues to travel in Real-Space.

According to the quantum mechanics a passage through a narrow region makes the future position uncertain. From the uncertainty principle, we have

$$\Delta P \cdot \Delta x = \hbar \, , \tag{H.12}$$

where ΔP is the uncertainty in the momentum and Δx is the uncertainty in the position.

From Eq.(H.12), we get

$$\Delta V = \hbar / (m \cdot \Delta x) \, , \tag{H.13}$$

where ΔV is the velocity uncertainty and m is the mass of particle.

We can say that a passage through a narrow region Δx causes a velocity uncertainty ΔV whose size is Eq.(H.13). Let us now apply the Planck unit to above equation.

Substituting the Planck mass and Planck length as thickness of barrier for Eq. (H.13), we get

$$\Delta V = \hbar / (m_{PL} \cdot L_{PL}) = \hbar / [(\hbar c / G)^{1/2} \cdot (G\hbar / c^3)^{1/2}] = \hbar / (\hbar / c) = c . \tag{H.14}$$

Eq.(H.14) indicates that even if the velocity of many-particle systems, i.e. spaceship is less than the velocity of light, the spaceship may achieve the velocity of light c by passing through the potential barrier. Therefore, the spaceship may keep the velocity of light c in Hyper-Space. A range of velocity V_S is c.

Hence, we may also consider that even if the spaceship is at rest, by turning on a fine-grained structure technology, the spaceship can plunge into Hyper-Space with the velocity of light. Therefore, it may not be necessary to accelerate the spaceship to a velocity of near c.

The time of passing through the potential barrier, from Eq.(H.4), we get

$$t = L_{PL} / c = (G\hbar / c^3)^{1/2} / c = (G\hbar / c^5)^{1/2} = t_{PL} . \tag{H.15}$$

Namely, the passing time is Planck time itself.

In addition, we have another form of uncertainty principle, that is

$$\Delta E \cdot \Delta t = \hbar , \tag{H.16}$$

where ΔE is the uncertainty of energy and Δt is the uncertainty of time.

Substituting of Planck time into Eq.(H.16) gives $\Delta E = 1.9 \times 10^9$ J. This value is Planck energy E_{PL}.

If above huge energy can be derived from passing through the potential barrier, we can avoid the following difficult problem:

1)   The mass of any object would become infinite ($m \rightarrow \infty$) at near the velocity of light and the structure of spaceship or crew would be broken.

2)   How can we get the power source of vast energy to accelerate any object to velocity near that of light.

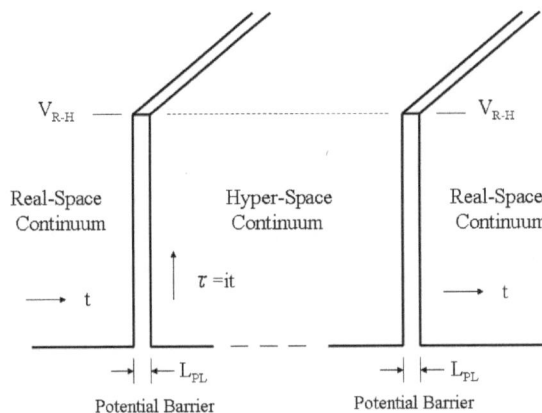

Figure H2: Properties of Hyper-Space

Finally, let us supplement the properties of Hyper-Space with a few more words on referring to Fig.H2. The Real-Space offered by Minkowski metric and Hyper-Space offered by Euclidean metric coexist, that is, the parallel space exists. And each space is isolated by potential barrier. The fracture of continuity of space means the crush of this potential barrier. Although the Hyper-Space is the space which has fractured the continuity of Real-Space, this cleft space shall also be a continuum. One and only difference is either real time or imaginary time.

H.4 WAVE FUNCTION OF SPACESHIP BY PATH INTEGRALS

The quantum tunneling is the quantum effect that the matter passes through the inaccessible region by its wave function. Forming a fine-grained structure as many-particle systems implies the matter wave. In quantum mechanics, since the wave function specifies the state of system, we consider here the wave function of spaceship by using the path integral approach. On referring to Fig.H3, $\phi(x_{a1}, \cdots x_{aN}, t_a)$ is the wave function of many-particle systems of N particles when the spaceship is formed a fine-grained structure at the point a. The wave function $\phi(x_{b1}, \cdots x_{bN}, it_b)$ of many-particle systems after passing through the potential barrier is given by using the path integral expression;

$$\phi(x_{b1}, \cdots x_{bN}, it_b) = \int_{-\infty}^{+\infty} [dx_{aN}] K(x_{b1}, \cdots x_{bN}, it_b; x_{a1}, \cdots x_{aN}, t_a) \phi(x_{a1}, \cdots x_{aN}, t_a), \tag{H.17}$$

where $\int [dx_{aN}] = \int \cdots \int dx_{a1} dx_{a2} \cdots dx_{aN}$.

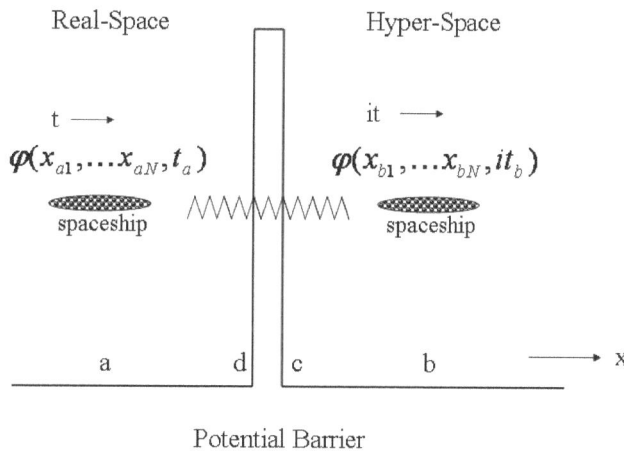

Figure H3: Wave function of Spaceship

Let d and c represent the position of potential barrier and let t_d and it_c(imaginary time) be the time of position d and c. The total amplitude which goes from the point in space-time (x_a, t_a) to (x_b, it_b), i.e. Feynman Kernel $K(b,a)$ is given by

$$K(b,a) = \int \int [dx_{cN}][dx_{dN}] K(b,c) K(c,d) K(d,a). \tag{H.18}$$

All paths between Real-Space and Hyper-Space are divided into two parts.

The time is real time between a and d and imaginary time between c and b. Finally, as point d comes closer and closer to point c, the real time t gets closer and closer to imaginary time it, i.e. analytic continuation.

Each kernel is represented as follows:

$$K(d,a) = K(x_{d1}, \cdots x_{dN}, t_d; x_{a1}, \cdots x_{aN}, t_a) = \int_a^d dx \cdot \exp[i/\hbar \cdot \int_{ta}^{td} dt L(\dot{x}, x, t)],$$

$$K(b,c) = K(x_{b1}, \cdots x_{bN}, it_b; x_{c1}, \cdots x_{cN}, it_c) = \int_c^b dx \cdot \exp[-1/\hbar \cdot \int_{tc}^{tb} dt L(\dot{x}, x, it)] \,, \tag{H19}$$

where L is the Lagrangian for system.

In the case of spaceship, the Lagrangian is a free particle system and is given by

$$L = \sum 1/2 \cdot m_{PL} \dot{x}_N^2 \,. \tag{H.20}$$

Taking the limit as (d-c) approaches zero, we get

$$K(b,a) = \int [dx_{d \to c, N}] K(b,c) K(d \to c, a) \,,$$

$$\phi(x_{b1}, \cdots x_{bN}, it_b) = \int_{-\infty}^{+\infty} [dx_{aN}] K(b,a) \phi(x_{a1}, \cdots x_{aN}, t_a) \,. \tag{H.21}$$

The wave function of spaceship is to be found out as above. To find out the kernel is equal to solve the following Schrödinger equation

$$H\phi(r_1, \cdots r_N) = i\hbar \frac{\partial \phi(r_1, \cdots r_N)}{\partial t} \,, \tag{H.22}$$

where H is the Hamiltonian operator.

The wave function of spaceship in Hyper-Space can be represented by the wave function in Real-Space.

APPENDIX.I

DERIVATION OF ELECTROGRAVITIC EQUATION

The electrogravitic equation similar to the formula from Weyl-Majumadar-Papapetrou solutions of the general relativity theory obtained by B.Ivanov can be derived shows as follows;

Weak field approximation of Einstein's General Relativity leads to the generalized formula of Lorenz force given by

$$F = q(E + v \times B) + m(E_g + v \times B_g) \,, \tag{I.1}$$

where q is the charge of the particle, m is the mass of the particle, E is the electric field, B is the magnetic field, v is the velocity of the particle, E_g is the electrogravitic field and B_g is the magnetogravitic field.

From which, gravitoelectric-electric coupling inside the static atom under high electric potential field becomes

$$qE + mE_g = 0 \,, \tag{I.2}$$

by assuming that the internal volume of an elementary particle is a region of force-free field like a superconductor. Then the gravitational field generated at the center of the charged particle by an external electric field becomes

$$E_g = -(q/m)E \,. \tag{I.3}$$

Comparing q/m values of an electron and a pion, E_g can be generated by an electron rather than a pion, hence we can let $q \approx e$ and $m \approx m_e$, where e is a charge of an electron and m_e is its mass. For the estimation of gravitational effect, we introduce the following approximation of the electrogravitic potential (Newton's serpentine) given by

$$\varphi_g = -\frac{eE}{m_e}\left(\frac{\delta^2 x}{\delta^2 + x^2}\right), \qquad \text{(I.4)}$$

which satisfies following conditions:

$$\partial \varphi_g(0)/\partial x = -(e/m_e)E, \qquad\qquad\qquad \text{(I.5-1)}$$

$$\partial \varphi_g(\pm\infty)/\partial x = 0, \qquad\qquad\qquad \text{(I.5-2)}$$

where δ is a length of the domain at which the new gravitational field is generated.

Figure.I1 shows the asymmetric electron orbit generated by the external electric field (A) and the shape of the electrogravitic potential induced by the deformation of the space generated at the center of the elementary particle (B).

By the electrogravitic potential at the center of the atomic nucleus becomes

$$\varphi_g = \varphi_g(r+\lambda) + \varphi_g(r-\lambda), \qquad\qquad\qquad \text{(I.6)}$$

where λ is a displacement of charge by an applied electric field and r is an orbital radius of the electron around the nucleus.

From which, the electrogravitic force generated by electrons circulating around the nucleus, the number of them equals the atomic number Z, is given by

$$E_g = \partial \varphi_g/\partial x = \delta^2 \frac{eE}{m_e}\sum_{i=1}^{Z}[(r_i+\lambda)^{-2}+(r_i-\lambda)^{-2}], \qquad\qquad \text{(I.7)}$$

when satisfying $|r_i \pm \lambda| \gg \delta$, where r_i is the orbital radius of each electron around the atomic nucleus. For simplifying the problem, we set $r_i \approx r_0$, then Eq.(I.7) becomes

$$E_g \approx Z\delta^2\frac{eE}{m_e}[(r_0+\lambda)^{-2}+(r_0-\lambda)^{-2}]. \qquad\qquad \text{(I.8)}$$

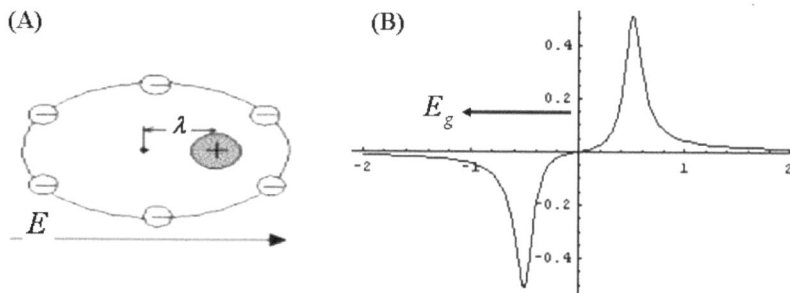

(A)   (B)

Figure I1: Electroravitic potential of a deformed atom

For relative lower voltage, we can suppose that $r_0 \gg \lambda$. Consequently Eq.(I.8) can be approximated as

$$E_g \approx -2\delta^2 Ze / (m_e r_0^2)\{1 + 3\lambda^2 / r_0^2 + \cdots\}E \ . \tag{I.9}$$

If we suppose that additional equivalent mass in a space due to the electric field[6] is canceled by negative mass created by the new gravitational field, we have the following formula given by

$$\int_V 2\pi G\varepsilon E^2 / c^2 dv - \int_V E_g^2 / (2c^2) dv = 0 \ , \tag{I.10}$$

where ε is permittivity, G is a gravitational constant and c is a light speed. From which, we have

$$\delta \approx \sqrt[4]{\pi\varepsilon G m_e^2 r_0^4 / e^2} \ . \tag{I.11}$$

If r_0 is replaced by Bohr's radius and ε equals ε_0, which is the permittivity of vacuum, we obtain the length of a domain where the new gravitational field is generated to be $\delta \approx 8.3 \times 10^{-22}$ (m), that is much smaller than the radius of electron, which is about 2.8×10^{-15} (m). Then, the acceleration of a dielectric material induced by high potential electric field can be approximated from Eqs.(I.9) and (I.11) as

$$E_g \approx -Z\sqrt{4\pi\varepsilon_r\varepsilon_0 G} \cdot E \ , \tag{I.12}$$

where ε_r is the specific inductive capacity of the dielectric material determined as $\varepsilon = \varepsilon_r\varepsilon_0$.

From which, the force generated by high potential electric field becomes

$$F \approx 8.62 \times 10^{-11} Z\mu S\sqrt{\varepsilon_r} \cdot V / t \ , \tag{I.13}$$

where μ is a total mass of the dielectric material per unit area, S is an area of the capacitor, t is the thickness of the capacitor and V is an impressed voltage.

T.T.Brown manufactured an electrogravitic effect as shown in Fig.I2 and he reported a large thrust generated which was associated with a spark between electrodes on the disc structure.

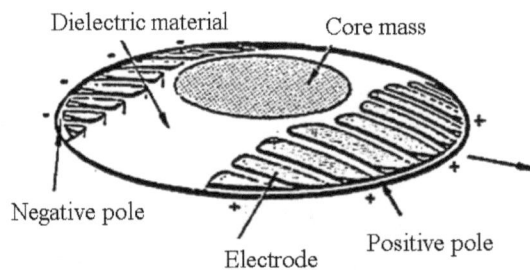

Figure I2: Electrogravitic craft proposed by T.T.Brown

APPENDIX.J

INTRODUCTION OF SQUEEZED LIGHT

According to the Quantum Optics, it is well known that the quantized field fluctuates; the zero-point energy and the vacuum fluctuations actually present severe problems in quantum field theory. The zero-point vacuum fluctuations actually give rise to observable effects, such as spontaneous emission of light, the Lamb shift and the Casimir effect. Although the light generated by laser is in a coherent state and its fluctuation (i.e., quantum noise) is exactly the same as for the vacuum, the squeezed light has less quantum noise in one of the quadratures than for a coherent state or a vacuum state; the fluctuations in that quadrature are squeezed. In this sense, squeezed light contains phase-dependent noise, reduced below that of the vacuum for some phases and enhanced above that of the vacuum. These squeezed states of light have been extensively investigated recently in quantum optics and have been experimentally realized also in Japan. To mathematically generate squeezed light, it is through the action of a "squeeze" operator defined as:

$$\hat{S}(\zeta) = \exp\left[\frac{1}{2}\zeta^* \hat{a}^2 - \frac{1}{2}\zeta(\hat{a}^+)^2\right], \quad \zeta = re^{i\theta},$$

(J.1)

where r is known as the squeeze parameter and $0 \le r < \infty$ and $0 \le \theta \le 2\pi$. When $|\zeta| = r = 0$, the squeezed state reduces to coherent states. Here, $\hat{a}$ and $\hat{a}^+$ are the annihilation and creation operators.

The squeeze operator $\hat{S}(\zeta)$ acting on the vacuum state (denoted as $|0\rangle$) would create the squeezed vacuum state (denoted as $|0,\zeta\rangle$):

$$|0,\zeta\rangle = \hat{S}(\zeta)|0\rangle$$

(J.2)

A more general squeezed state for a single mode can be obtained by applying the displacement operator $\hat{D}(\alpha)$ as

$$|\alpha,\zeta\rangle = \hat{D}(\alpha)\hat{S}(\zeta)|0\rangle, \quad \alpha = |\alpha|e^{i\theta}$$

(J.3)

that is, the squeezed state $|\alpha,\zeta\rangle$ is obtained by first squeezing the vacuum and the displacing it. By displacing the vacuum state $|0\rangle$, the coherent states are given as $|\alpha\rangle = \hat{D}(\alpha)|0\rangle$.

When $\alpha=0$, it is known as a squeezed vacuum state. Such a state is not the vacuum state so long as $\zeta \neq 0$, but rather a superposition of states containing even numbers of particles. The variances of two quadratures of the squeezed vacuum state for $\theta=0$ are given by following Eq.(J.4).

For $\theta=0$, evidently squeezing exists in the $\hat{X}_1$ quadrature. For $\theta=\pi$, the squeezing will appear in the $\hat{X}_2$ quadrature.

$$\left\langle (\Delta \hat{X}_1)^2 \right\rangle_{sq-vac} = \frac{1}{4}e^{-2r}, \quad \left\langle (\Delta \hat{X}_2)^2 \right\rangle_{sq-vac} = \frac{1}{4}e^{2r}.$$

(J.4)

A representation of the squeezed vacuum state for $\theta=0$, where the fluctuations in X_1 are reduced, is given in Fig.J1 (a), whereas for $\theta=\pi$, where the fluctuations in X_2 are reduced, is given in Fig.J1 (b).

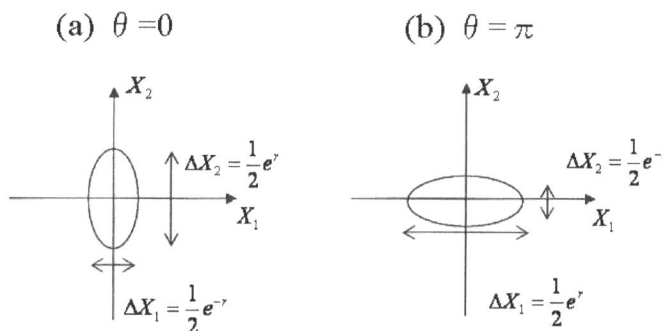

Figure J1: Error ellipse for a Squeezed vacuum state

(a) Coherent State   (b) Vacuum State

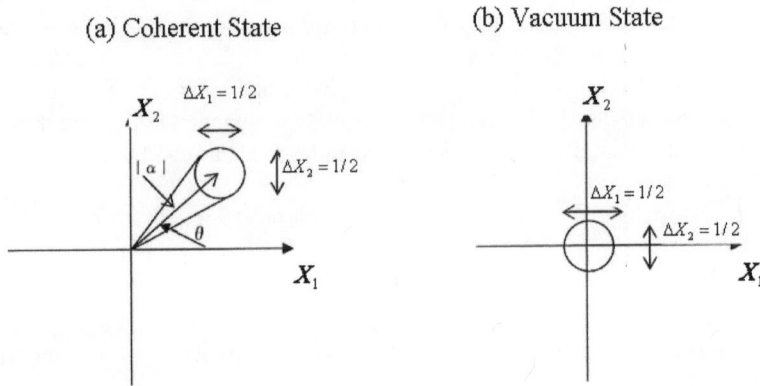

Figure J2: Phase-Space portrait (Non-squeezed state)

In the cases with $\theta=0$ or $\theta=\pi$, the squeezing is along either $\hat{X}_1$ or $\hat{X}_2$. A "squeezed state" is a state whose fluctuations in one quadrature phase are less than zero-point fluctuations (or the fluctuations in any coherent state), and whose fluctuations in the other phase are larger than zero-point fluctuations. Namely, the squeeze operator attenuates one component of the amplitude, and it amplifies the other component. The degree of attenuation and amplification is determined by squeeze parameter $r=|\zeta|$. A coherent state and a vacuum state where in both cases the fluctuations of the quadrature operators are equal, $\Delta X_1 = \Delta X_2 = 1/2$ (see Fig.J2, (a),(b)).

As is shown above, the squeezed state $|\alpha,\zeta\rangle$ has the same expected complex amplitude as the corresponding coherent state $|\alpha\rangle$, and it is a minimum-uncertainty state for X_1 and X_2.

The difference lies in its unequal uncertainties for X_1 and X_2. That is, squeezed states are less noisy (or larger noisy) than for a field in a vacuum state. This implies that ZPF noise (quantum noise) can be manipulated.

For a reference, the graph of squeezed vacuum state of electric field versus time is shown as Fig.J3. Although the fluctuation in coherent state is temporally constant, the fluctuation in squeezed state periodically varies from the maximum value of εe^{r} to the minimum value of εe^{-r}.

Here, $\varepsilon = (\hbar\omega/2\varepsilon_0 V)^{1/2}$, ε_0 is dielectric constant of vacuum, $\hbar$ is Planck Constant (1.054×10^{-34} J.s), ω is angle frequency (rad/s) and V is volume (m³).

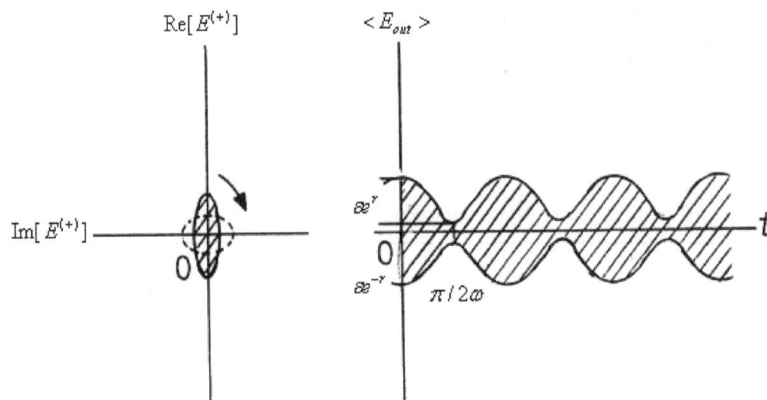

Figure J3: Electron field for Squeezed vacuum state (Matsuoka, 2000)

The graphs of electric field versus time for three states of the electromagnetic field are shown as Fig.J4 ($\alpha \neq 0$). For a coherent state, the rotation of the error circle leads to a constant value for the variance of the electric field (Fig.J4, (a)). For a squeezed state, the rotation of the error ellipse leads to a variance that oscillates with frequency 2ω (Fig. J4, (b),(c)).

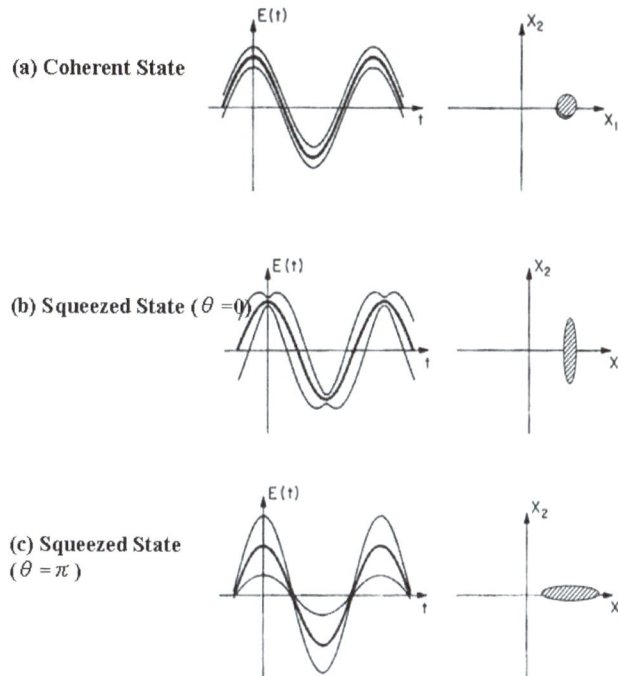

Figure J4: Electron field for Coherent & Squeezed states

Figure.J5 shows the squeezed vacuum state. Vacuum state line (: ΔE ; pink) indicates, non-squeezed usual vacuum state (coherent laser beam), and the fluctuations in the electric field ΔE is constant at all times. The fluctuations in the electric field ΔE are the same value both interior of laser beam and exterior of laser beam. The fluctuations in the electric field ΔE exterior of laser beam are usual vacuum state. On the contrary, concerning the squeezed vacuum state induced by squeezed light, the fluctuation in squeezed state periodically varies from the maximum value of εe^{r} to the minimum value of εe^{-r} ; $\varepsilon = (\hbar\omega / 2\varepsilon_0 V)^{1/2}$ (see Fig.J5 (a)).

Next, let us consider about vacuum energy density shown in Fig.J5, (b).

As mentioned previously, non-squeezed vacuum energy density ($\varepsilon_0 \varepsilon^2$) is constant like $\varepsilon_0 \varepsilon^2 = \varepsilon_0 (\sqrt{\hbar\omega / 2\varepsilon_0 V})^2 = \hbar\omega / 2V \, (J / m^3)$. In the case of large value of squeeze parameter "r", the energy density of vacuum increases exponentially in accordance with $\varepsilon_0 \varepsilon^2 e^{2r} = \hbar\omega e^{2r} / 2V \, (J / m^3)$. The squeezed vacuum exhibits larger fluctuations, and hence more energy density, than the vacuum in space-time regions. On the other hand, the squeezed vacuum also exhibits smaller fluctuations, and hence less energy density, than the vacuum in space-time regions. Consequently, the average vacuum energy density in squeezed state (blue line) increases as compared to non-squeezed vacuum state (pink line). Thus, the squeezed light can alter the energy density of vacuum, that is, vacuum perturbation (ZPF) can be controlled by squeezed light technology.

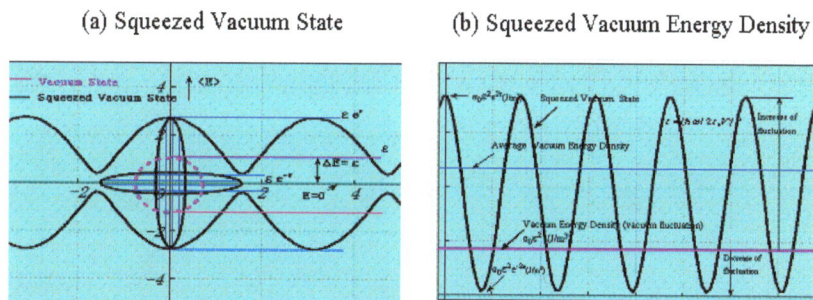

Figure J5: Electron field for Squeezed vacuum state

APPENDIX.K

ENTROPY OF R-N BLACK HOLE

From compactifying D-p branes, reducing the space-time dimensions to d-dimensions, the solution in Einstein frame is given by

$$ds_d^2 = -\lambda^{-(\frac{d-3}{d-2})} dt^2 + \lambda^{\frac{1}{d-2}} d\tilde{x}_{d-1}^2 .$$

(K.1)

Setting $d=5$, we have

$$ds_5 = -\lambda^{-2/3} dt^2 + \lambda^{1/3} d\tilde{x}_4^2 = -\lambda^{-2/3} dt^2 + \lambda^{1/3} (dx_1^2 + dx_2^2 + dx_3^2 + dx_4^2)$$
$$= -\lambda^{-2/3} dt^2 + \lambda^{1/3} (dr^2 + r^2 d\Omega_3^2)$$

(K.2)

The five dimensional line element is given from Eq.(K.2) as follows:

$$ds^2 = -\lambda^{-2/3} dt^2 + \lambda^{1/3} (dr^2 + r^2 d\Omega_3^2) ,$$

(K.3)

and λ is given by

$$\lambda = (1 + \frac{c_1 Q_1}{r^2})(1 + \frac{c_5 Q_5}{r^2})(1 + \frac{c_n n}{r^2}) .$$

(K.4)

The coefficients in Eq.(K.4)

$$c_1 = \frac{4 G_5 R}{g_s \pi \alpha'}, \quad c_5 = \alpha' g_s, \quad c_n = \frac{4 G_5}{\pi R} ,$$

(K.5)

are defined so that the charges Q_1, Q_5 and n are naturally integer quantized.

In these expressions, g_s is the string coupling constant, G_5 is the five-dimensional Newton's constant, α' is given by the string tension $T = 1/2\pi l_s^2 = 1/2\pi\alpha'$.

Eq.(K.5) yields

$$c_1 c_5 c_n = 16 G_5^2 / \pi^2 .$$

(K.6)

As a special case of solution, when the charges are chosen such that

$$c_1 Q_1 = c_5 Q_5 = c_n n = r_0^2 ,$$

(K.7)

Eq.(K.3) results in five-dimensional extremal Reissner-Nord ström black hole solution

$$ds^2 = -(1 - \frac{r_0^2}{r^2})^2 dt^2 + \frac{dr^2}{(1 - \frac{r_0^2}{r^2})^2} + r^2 d\Omega_3^2 .$$

(K.8)

From Eq.(K.7), following equations

$$c_1 = \frac{r_0^2}{Q_1}, \ c_5 = \frac{r_0^2}{Q_5}, \ c_n = \frac{r_0^2}{n},$$

(K.9)

are obtained. Substituting these equations into Eq.(K.6), we get the following:

$$c_1 c_5 c_n = \frac{r_0^6}{Q_1 Q_5 n} = \frac{16 G_5^2}{\pi^2} \ ,$$

(K.10)

then

$$G_5 = \frac{\pi r_0^3}{4\sqrt{Q_1 Q_5 n}} \ .$$

(K.11)

Since the horizon area A is $2\pi^2 r_0^3$, the black hole entropy S_{BH} is obtained as

$$S_{BH} = \frac{A}{4G_5} = \frac{2\pi^2 r_0^3}{4 \cdot \frac{\pi r_0^3}{4\sqrt{Q_1 Q_5 n}}} = 2\pi\sqrt{Q_1 Q_5 n} \ .$$

(K.12)

Index

A

B

C

D

T

tachyon 57, 93-94
tachyon tunneling 93

U

Urashima effect 5
Unruh effect 58

V

vacuum energy 28-29, 41-42, 47-51, 115, 127
vacuum field 21-22, 80, 105
vacuum-fluctuation battery 31, 115
vacuum perturbation 23, 78-79, 81, 143
vacuum polarization 31, 58, 102
vector potential 27, 65-68, 76-77, 96-98, 100

W

warp bubble 56-58, 89
warp drive 5, 35, 56-58, 84-85
warp navigation 84
Weyl-Majumandar-Papapetrou solutions 97, 138
Weyl field 97
wormhole 5, 57, 84-90, 93
wormhole engineering 84-85

Y

Yang-Mills field 34

Z

zero-point field 21, 31, 64, 66, 75-76, 78, 81
zero-point energy 5-6, 21, 28-31, 75, 78, 81, 140
ZPF 21, 52, 55, 63-66, 75-78, 143
ZPF field 52, 63-64, 66-68, 71, 98, 115
ZPF propulsion system 73, 75, 114

www.ingramcontent.com/pod-product-compliance
Lightning Source LLC
Chambersburg PA
CBHW080020240326

41598CB00075B/611